Selected Papers from the 9th Symposium on Micro-Nano Science and Technology on Micromachines

Selected Papers from the 9th Symposium on Micro-Nano Science and Technology on Micromachines

Special Issue Editors

Norihisa Miki
Koji Miyazaki
Yuya Morimoto

MDPI • Basel • Beijing • Wuhan • Barcelona • Belgrade

MDPI

Special Issue Editors

Norihisa Miki
Keio University
Japan

Koji Miyazaki
Kyushu Institute of Technology
Japan

Yuya Morimoto
The University of Tokyo
Japan

Editorial Office
MDPI
St. Alban-Anlage 66
4052 Basel, Switzerland

This is a reprint of articles from the Special Issue published online in the open access journal *Micromachines* (ISSN 2072-666X) from 2018 to 2019 (available at: https://www.mdpi.com/journal/micromachines/special_issues/MNST2018).

For citation purposes, cite each article independently as indicated on the article page online and as indicated below:

LastName, A.A.; LastName, B.B.; LastName, C.C. Article Title. *Journal Name* **Year**, *Article Number*, Page Range.

ISBN 978-3-03921-696-3 (Pbk)
ISBN 978-3-03921-697-0 (PDF)

Contents

About the Special Issue Editors

Norihisa Miki received his Ph.D. in mechano-informatics from University of Tokyo in 2001. He then worked at the MIT microengine project as a postdoctoral associate and later as a research engineer. He joined the Department of Mechanical Engineering of Keio University in 2004 as an associate professor and became a full professor in 2017. His research interests include micro/nano biomedical devices and information communication technologies (ICT). He was a JST PRESTO researcher from 2010 to 2016 and at the Kanagawa Institute of Industrial Science and Technology (formerly, Kanagawa Academy of Science and Technology). He was the general chair of the JSME 8th and 9th Symposia on Micro Nano Science and Technology in 2017 and 2018. He co-founded a healthcare startup, LTaste Inc., in 2017.

Koji Miyazaki received his Ph.D. in mechanical engineering and science from Tokyo Institute of Technology in 1999. He joined the Department of Mechanical and Control Engineering at Kyushu Institute of Technology in 1999 as a lecturer and became a full professor in 2011. His research interests include the thermophysical properties of nano-structured materials and micro thermal devices such as a thermoelectric micro-generators. He stayed at UCLA from 2000 to 2001 and at MIT from 2001 to 2002 as a visiting scholar. He was a JST PRESTO researcher from 2004 to 2008, and he is currently a PI of JST CREST, since 2017. He will be a general chair of the JSME 10th Symposium on Micro Nano Science and Technology in 2019.

Yuya Morimoto received his M.E. degree from the University of Tokyo in 2009. Between 2009 and 2011, he worked at Fujifilm Corporation on the R&D of medical endoscopes. He received his Ph.D. in mechano-informatics from the University of Tokyo in 2014 and then joined the Institute of Industrial Science (IIS), University of Tokyo as an assistant professor. He is currently an associate professor at the Department of Mechano-Informatics, Graduate School of Information Science and Technology, University of Tokyo. His research interests are biohybrid robotics and biofabrication with microengineering techniques. He was a committee member of JSME 8th and 9th symposia on Micro Nano Science and Technology in 2017 and 2018.

Editorial

Editorial for the Special Issue of Selected Papers from the 9th Symposium on Micro-Nano Science and Technology on Micromachines

Norihisa Miki [1,*], Koji Miyazaki [2] and Yuya Morimoto [3,4]

1 Department of Mechanical Engineering, Keio University, 3-14-1 Hiyoshi, Kohoku-ku, Yokohama, Kanagawa 223-8522, Japan
2 Department of Mechanical and Control Engineering, Kyushu Institute of Technology, 1-1 Sensui-cho, Tobata-ku, Kitakyushu, Fukuoka 804-8550, Japan; miyazaki@mech.kyutech.ac.jp
3 Department of Mechano-Informatics, Graduate School of Information, Science and Technology, The University of Tokyo, 7-3-1 Hongo, Bunkyo-ku, Tokyo 113-8656, Japan; y-morimo@hybrid.t.u-tokyo.ac.jp
4 Institute of Industrial Science (IIS), The University of Tokyo, 4-6-1 Komaba, Meguro-ku, Tokyo 153-8505, Japan
* Correspondence: miki@mech.keio.ac.jp; Tel.: +81-45-566-1430

Received: 12 September 2019; Accepted: 12 September 2019; Published: 17 September 2019

The Micro-Nano Science and Technology Division of the JSME (Japan Society of Mechanical Engineers) promotes academic activities to pioneer novel research topics on microscopic mechanics. The division encourages interdisciplinary studies to deeply understand physical/chemical/biological phenomena at the micro/nano scale and to develop applied technologies. Since 2009, the past seven symposiums on Micro-Nano Science and Technology have taken place in a more interdisciplinary manner, incorporating the related societies of electronics and applied physics. We have promoted in-depth studies and interactions between researchers/engineers in various fields with more than 140 papers presented at each symposium for the past few years. Thanks to the previous activities and the great effort of the committee members, the Micro-Nano Science and Technology Division has been recognized as a formal division within the JSME.

This Special Issue collects 14 papers from the 9th Symposium on Micro-Nano Science and Technology, which was held from October 30 through 1 November 2018, in Sapporo, Hokkaido, Japan. All of the papers highlight new findings and technologies at micro/nano scales relating to a wide variety of fields of mechanical engineering, from fundamentals to applications.

This issue present new fabrication technologies ranging from nano, micro, and mili scales. Direct writing of copper (Cu) in an ambient environment using femtosecond laser was proposed [1]. The laser reduces a glyoxylic acid Cu complex, which can be spin-coated onto a glass substrate. The resulting resistance of the patterned Cu was found to be large. The authors carefully investigated it and found the re-oxidation of the glyoxylic acid Cu complex to be the source. Nano-scale surface modification is known to be effective for control of heat transfer. Given the difficulty of direct observation of the phenomena, molecular dynamics simulation was conducted, which nicely explained the contact angle and water condensation at the surface [2]. Micro/nano fabrication is not limited to inorganic material but organic material that is soft, flexible, and biocompatible. Printing of a stimuli-responsive hydrogel, which includes printing an N-isopropylacrylamide-based stimuli-responsive pre-gel solution and an acrylamide-based non-responsive pre-gel solution in a supporting viscous liquid, and polymerizing the printed structures using ultraviolet (UV) light irradiation, was introduced [3]. Not only do the fabrication processes enable three-dimensional structures but the formed hydrogel can also respond to the stimuli. The authors claimed the process as 4D printing. Kirigami structures can generate large deformation with good controllability while the manufacturing process is rather two-dimensional and compatible with micro/nano technologies. However, the edges of the structures are typically not well constrained and cause instability in the motion. Therefore, a model comprising of connected springs in series with different rigidities in the regions close to the ends and the center is proposed [4]. It showed good agreement with experiments and will contribute to the theoretical design of kirigami structures.

Fabricated micro/nano features and devices must be assembled and packaged at the mili-scale to exhibit the best performance. The contact resistance when the electronic components are mounted using elastic adhesives was investigated, which is crucial in solderless writing in low temperature at low cost [5]. The careful investigation with respect to the contact pressure and Cu layer thickness led to the development of the sandwich structure to decrease the contact resistance. Micro/nano medical devices that exploit the small size and beneficial scale effects have been developed, however, the connection to the body is by far the most challenging. The connecting mechanism between the artificial blood vessels to facilitate the surgical procedure was proposed and demonstrated [6]. The mechanism allows blood to have contact only with the highly biocompatible surface; that is, the inner surface of the artificial blood vessels. The biocompatibility was experimentally investigated.

Sensors are one of the major applications of micro/nano technologies, which exploit beneficial scale effects in electro/magneto/mechanical science and engineering. A near-infrared spectrometer with a wide wavelength range using a plasmonic gold grating was proposed and demonstrated [7]. By improving the spectrum derivation procedure, the wavelength range covers 1200 to 1600 nm. A thin-film magnetic field sensor with a logarithmic amplifier was newly proposed [8]. The amplifier can translate hundreds of MHz signals to a direct current (DC) voltage signal which is proportional to the radio frequency (RF) signal. A whole sensor system can be small enough to be practically used to detect foreign materials in industrial and medical products. Tactile sensation is considered to be the next tool for the intuitive and efficient human/computer interface. A thermal tactile sensation display, which controls the effective thermal conductivity, was proposed and demonstrated [9]. A highly thermally conductive liquid metal is introduced into the device, whose amount controls the effective thermal conductivity of the device. The range of the effective thermal conductivity was experimentally deduced and human perception tests were conducted to verify the concept.

Micro/Nano fluidics have been studied from their fundamentals to their biomedical applications. This Special Issue covers these topics with five papers. First, separation of nano- and micro-particle flows in branched microfluidic channels using thermophoresis [10]. Localized temperature increases near the branch are achieved using the Joule heat from a thin-film micro electrode embedded in the bottom wall of the microfluidic channel. The particle flow into one of the outlets is blocked by microscale thermophoresis since the particles are repelled from the hot region in the experimental conditions used here. The nano-particle case was also discussed theoretically and experimentally. The steady streaming that can generate net mass flow from zero-mean vibration is attracting many researchers in this field. To achieve the steady streaming, the numerical analysis for three-dimensional and unsteady flow was proposed [11]. The particle trajectories induced around a cylindrical micro-pillar under circular vibration was solved in the Lagrangian frame and the results were converted to a stationary Eulerian frame to compare with the experimental results, which showed good agreement. The proposed model can be a strong tool to design the micro scale flow of interest.

Biomedical applications using micro/nano fluidics and biocompatible polymer material, in particular hydrogels, are discussed. The degeneration of adipocyte has been reported to cause obesity, metabolic syndrome, and other diseases. To treat these diseases, an effective in vitro evaluation and drug-screening system for adipocyte culture is required. An in vitro three-dimensional cell culture system to enable the monitoring of lipid accumulation by measuring electrical impedance was proposed [12]. The relationship between the impedance and lipid accumulation of adipocytes was investigated experimentally and the lipid accumulation of adipocytes was found to be monitored in real time by the electrical impedance during in vitro culture. Reconstructing a three-dimensional muscle using living cells is promising for restoration of damaged muscles. However, the regenerated tissue exhibits a weak construction force due to the insufficient tissue maturation. A cell-laden core-shell hydrogel microfiber as a three-dimensional culture to control the cellular orientation with

Micromachines **2019**, *10*, 618

cyclic mechanical stimulation was proposed and demonstrated [13]. The directions of the myotubes were oriented and the mature myotubes could be successfully formed by cyclic stretch stimulation. An anchoring device with pillars to immobilize an adipocyte microfiber was proposed to track the specific positions of the microfiber for a long period [14]. Temporal observations of the microfiber on the device for a month successfully revealed the function and morphology of three-dimensional cultured adipocytes. Lipolysis of the microfiber's adipocytes by applying reagents with an anti-obesity effect was also demonstrated, which indicates the effectiveness of the system for drug tests.

We would like to thank all the contributing authors for their excellent research work. We appreciate all the reviewers who provided valuable comments to improve the quality of the papers and the tremendous support from the editorial staff of Micromachines.

References

1. Mizoshiri, M.; Aoyama, K.; Uetsuki, A.; Ohishi, T. Direct Writing of Copper Micropatterns Using Near-Infrared Femtosecond Laser-Pulse-Induced Reduction of Glyoxylic Acid Copper Complex. *Micromachines* **2019**, *10*, 401. [CrossRef] [PubMed]
2. Hiratsuka, M.; Emoto, M.; Konno, A.; Ito, S. Ito Molecular Dynamics Simulation of the Influence of Nanoscale Structure on Water Wetting and Condensation. *Micromachines* **2019**, *10*, 587. [CrossRef] [PubMed]
3. Onoe, H.; Uchida, T. 4D Printing of Multi-Hydrogels Using Direct Ink Writing in a Supporting Viscous Liquid. *Micromachines* **2019**, *10*, 433. [CrossRef] [PubMed]
4. Taniyama, H.; Iwase, E. Design of Rigidity and Breaking Strain for a Kirigami Structure with Non-Uniform Deformed Regions. *Micromachines* **2019**, *10*, 395. [CrossRef] [PubMed]
5. Sato, T.; Koshi, T.; Iwase, E. Resistance Change Mechanism of Electronic Component Mounting through Contact Pressure Using Elastic Adhesive. *Micromachines* **2019**, *10*, 396. [CrossRef] [PubMed]
6. Watanabe, A.; Miki, N. Connecting Mechanism for Artificial Blood Vessels with High Biocompatibility. *Micromachines* **2019**, *10*, 429. [CrossRef] [PubMed]
7. Suido, Y.; Yamamoto, Y.; Thomas, G.; Ajiki, Y.; Kan, T. Extension of the Measurable Wavelength Range for a Near-Infrared Spectrometer Using a Plasmonic Au Grating on a Si Substrate. *Micromachines* **2019**, *10*, 403. [CrossRef] [PubMed]
8. Nakai, T. Magneto-Impedance Sensor Driven by 400 MHz Logarithmic Amplifier. *Micromachines* **2019**, *10*, 355. [CrossRef] [PubMed]
9. Hirai, S.; Miki, N. A Thermal Tactile Sensation Display with Controllable Thermal Conductivity. *Micromachines* **2019**, *10*, 359. [CrossRef] [PubMed]
10. Tsuji, T.; Matsumoto, Y.; Kugimiya, R.; Doi, K.; Kawano, S. Separation of Nano- and Microparticle Flows Using Thermophoresis in Branched Microfluidic Channels. *Micromachines* **2019**, *10*, 321. [CrossRef] [PubMed]
11. Kaneko, K.; Osawa, T.; Kametani, Y.; Hayakawa, T.; Hasegawa, Y.; Suzuki, H. Numerical and Experimental Analyses of Three-Dimensional Unsteady Flow around a Micro-Pillar Subjected to Rotational Vibration. *Micromachines* **2018**, *9*, 668. [CrossRef] [PubMed]
12. Zemmyo, D.; Miyata, S. Evaluation of Lipid Accumulation Using Electrical Impedance Measurement under Three-Dimensional Culture Condition. *Micromachines* **2019**, *10*, 455. [CrossRef] [PubMed]
13. Bansai, S.; Morikura, T.; Onoe, H.; Miyata, S. Effect of Cyclic Stretch on Tissue Maturation in Myoblast-Laden Hydrogel Fibers. *Micromachines* **2019**, *10*, 399. [CrossRef] [PubMed]
14. Morimoto, Y.; Nishimura, K.; Yokomizo, A.; Takeuchi, S. Temporal Observation of Adipocyte Microfiber Using Anchoring Device. *Micromachines* **2019**, *10*, 358. [CrossRef] [PubMed]

![micromachines logo] *micromachines*

MDPI

Article

Direct Writing of Copper Micropatterns Using Near-Infrared Femtosecond Laser-Pulse-Induced Reduction of Glyoxylic Acid Copper Complex

Mizue Mizoshiri [1,*], Keiko Aoyama [2], Akira Uetsuki [3] and Tomoji Ohishi [3]

[1]   Department of Mechanical Engineering, Nagaoka University of Technology, Nagaoka 940-2188, Japan
[2]   Department of Mechanical and Aerospace Engineering, Nagoya University, Nagoya 464-8603, Japan; pixwhxbl.color@gmail.com
[3]   Department of Applied Chemistry, Shibaura Institute of Technology, Tokyo 135-8548, Japan; Mc18006@shibaura-it.ac.jp (A.U.); tooishi@sic.shibaura-it.ac.jp (T.O.)
*   Correspondence: mizoshiri@mech.nagaokaut.ac.jp; Tel.: +81-258-47-9765

Received: 14 May 2019; Accepted: 13 June 2019; Published: 17 June 2019

Abstract: We have fabricated Cu-based micropatterns in an ambient environment using femtosecond laser direct writing to reduce a glyoxylic acid Cu complex spin-coated onto a glass substrate. To do this, we scanned a train of focused femtosecond laser pulses over the complex film in air, following which the non-irradiated complex was removed by rinsing the substrates with ethanol. A minimum line width of 6.1 μm was obtained at a laser-pulse energy of 0.156 nJ and scanning speeds of 500 and 1000 μm/s. This line width is significantly smaller than that obtained in previous work using a CO_2 laser. In addition, the lines are electrically conducting. However, the minimum resistivity of the line pattern was 2.43×10^{-6} Ω·m, which is ~10 times greater than that of the pattern formed using the CO_2 laser. An X-ray diffraction analysis suggests that the balance between reduction and re-oxidation of the glyoxylic acid Cu complex determines the nature of the highly reduced Cu patterns in the ambient air.

Keywords: laser direct writing; femtosecond laser; glyoxylic acid Cu complex; reduction; Cu micropattern

---

1. Introduction

Laser direct writing of metal micropatterns has attracted attention from fields such as printed electronics and microelectromechanical systems. Two-dimensional (2D) metal micropatterns are generally fabricated using well-established methods of semiconductor technology consisting of lithography, metallic film deposition methods, and etching processes. However, deposition methods such as sputtering and evaporative coating must be done in an inert atmosphere, making it difficult to fabricate 2D metal micropatterns in air. In addition, multiple complicated steps such as lithography, metal deposition, and etching are needed to form metal micropatterns.

To overcome this problem, direct writing using laser-induced reduction has been proposed [1–4]. With this technology, Cu micropatterns are directly written using a laser-induced thermochemical reduction of copper oxide nanoparticles (NPs), such as CuO and Cu_2O NPs, which are mixed with reductants and dispersants, and reduced to Cu by laser irradiation. When a CuO NP solution containing CuO NPs, polyvinylpyrrolidone (PVP), and ethylene glycol (EG) is irradiated by continuous-wave and nanosecond-pulsed lasers, acetaldehyde generated by dehydrating EG reduces the CuO NPs to Cu NPs, which are subsequently sintered to form Cu micropatterns [1]. When using a Cu_2O NP solution, which contains Cu_2O NPs, 2-propanol, and PVP, 2-propanol and PVP react thermally to generate formic acid, which then reduces Cu_2O to Cu [2].

Two-dimensional Ni micropatterns can also be formed on glass and polyimide films using laser reductive sintering [3,4]. In this technique, NiO NPs mixed with toluene and α-terpineol are reduced to Ni by nanosecond-laser-induced thermochemical reduction. This technology has been used to fabricate Ni microwires with highly transparent electrodes on flexible films.

We have also fabricated Cu-based micropatterns using femtosecond-laser-reductive sintering of CuO NPs. An advantage of this approach is that the short pulse duration leads to rapid heating and cooling of the materials because the total irradiated energy can be reduced in femtosecond laser heating. In this research, irradiation by femtosecond laser pulses thermally reduces CuO NPs mixed with PVP and EG. Further, Cu- and Cu_2O-rich micropatterns can be formed selectively by tuning the laser-irradiation conditions. The temperature coefficients of resistance of the Cu- and Cu_2O-rich micropatterns are positive and negative, respectively, which is consistent with their respective metallic and semiconductive properties [5,6].

Metal complexes are promising candidate materials for direct laser writing using reduction, as demonstrated by the reduction of Cu complexes to produce 2D Cu micropatterns [7–9]. Typically, these Cu complexes are easily reduced at a relatively low temperature (~200 °C) [7]. In other work, Cu formate has been reduced to form Cu NPs by irradiation with an ultraviolet (UV) nanosecond-pulsed laser in an inert atmosphere under N_2 gas flow [7,8]. A glyoxylic acid Cu (GACu) complex has also been developed for ambient-air Cu micropatterning using a CO_2 laser [9]. This complex can be reduced in ambient air because of its high resistance to oxidation, ease of reduction, and strong absorption of CO_2-laser irradiation. The minimum line width was ~200 μm, and the resistivity of the resulting Cu micropattern was ~3×10^{-7} $\Omega \cdot$m. However, finer line patterning has not been achieved because the line width depends on the irradiated diameter of the CO_2 laser beam, which cannot be focused to a smaller spot diameter due to its long wavelength.

In this study, we report herein 2D Cu micropatterns fabricated in ambient air by using femtosecond laser reduction of a GACu complex to fabricate finer patterns with small line width. We first investigate the absorption properties of GACu, following which we discuss the patterning properties of GACu, such as resolution, crystal structure, and resistivity.

2. Experimental Methods

2.1. Direct Writing Process of Two-Dimensional Cu Micropatterns

Figure 1 shows schematically the process for direct writing of 2D Cu micropatterns. A GACu complex was prepared using a previously reported method [9]. First, glyoxylic acid (4.5 mmol, Sigma Aldrich, St. Louis, MO, USA) dissolved in H_2O (5 mL, FUJIFILM Wako Pure Chemical Corporation, Tokyo, Japan) was adjusted to pH 7 by adding NaOH aqueous solution (10 wt%, FUJIFILM Wako Pure Chemical Corporation). Next, $CuSO_4 \cdot 5H_2O$ (4.5 mmol, FUJIFILM Wako Pure Chemical Corporation) dissolved in 5 mL H_2O was added to the GA solution and stirred for three hours. The GACu complex precipitated from the solution and was filtered out, washed by H_2O, and dried in a cooled, reduced-pressure atmosphere.

The GACu complex (6.0 mmol) was dissolved into a 2-amino-ethanol: ethanol (1:2, 3 mL, FUJIFILM Wako Pure Chemical Corporation) solution and was then spin-coated onto a glass substrate. The spin-coated film was heated at 50 °C using a hot plate for 30 min. To accomplish laser direct writing, we used a commercially-available femtosecond laser direct writing system (Photonic Professional GT, Nanoscribe GmbH, Eggenstein-Leopolds-hafen, Germany) to form Cu micropatterns by reducing and precipitating the GACu complex.

The wavelength, pulse duration, and repetition frequency of the femtosecond laser were 780 nm, 120 fs, and 80 MHz, respectively. The laser pulses had a Gaussian intensity distribution and were focused onto the surface of GACu complex films using an objective lens with a numerical aperture (NA) of 0.75. The focal spot diameter was 1.3 μm. The sample substrates coated with the GACu complex film were scanned using an xyz-piezo stage. The maximum scanning speed was 1000 μm/s.

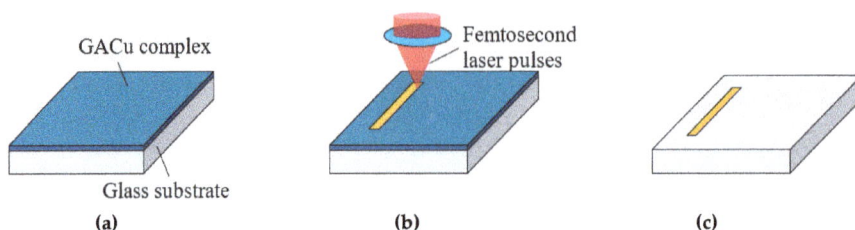

Figure 1. (a) Spin-coating of a glyoxylic acid Cu (GACu) complex film on a glass substrate. (b) Femtosecond-laser direct writing by reduction of the GACu complex film. (c) Nonirradiated GACu complex removed by rinsing the substrate with ethanol.

2.2. Evaluation of GACu Complex Films and Cu Micropatterns

The absorption properties of the GACu complex film are important for laser direct writing. The absorbance of the film in the UV-to-visible range was examined using a UV-visible spectrometer (UV-2600 100V JP, Shimadzu, Kyoto, Japan). The line width was measured using field-emission scanning electron microscopy (FE-SEM). The crystal structure of the micropatterns was examined using X-ray diffraction (XRD) (Rint Rapid-S diffractometer, Rigaku, Tokyo, Japan). The diameter of the collimated X-ray beam was 0.3 mm, and the angle of incidence was 20°.

The resistance of the line patterns was measured using a multimeter (Truevolt series 34465A, Keysight Technology, Santa Rosa, CA, USA). The resistivity was calculated from the resistance and the cross section of the line patterns which were obtained using a surface coder (SURFCODER ET200, Kosaka Laboratory Ltd., Tokyo, Japan).

3. Results and Discussion

Here we discuss the properties of the Cu micropatterns on the SiO_2 glass substrates. First, we examine the absorption of the GACu complex film, following which we investigate the properties of the micropatterns such as line width, the generation of Cu-based micropatterns, and their resistivities.

3.1. Absorption of the GACu Complex Film

Figure 2 shows the absorption spectrum of the GACu complex film on a glass substrate. The absorbance at 780 nm was almost the same as that at 390 nm, which indicates that single-photon absorption, rather than multi-photon absorption, is dominant during irradiation with femtosecond laser pulses at 780 nm. However, it is possible that three-photon absorption may occur. The relatively small absorption allows precise control of energy absorbed by the material by controlling the irradiated energy.

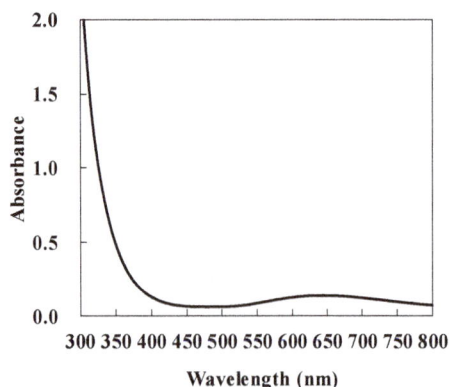

Figure 2. Absorption spectrum of GACu complex thin film.

3.2. Patterning Properties

We next examined the relationship between pattern line width and laser-irradiation conditions, such as pulse energy and scanning speed. Figure 3a shows how the line width depends on the pulse energy at scanning speeds of 300, 500, and 1000 μm/s. Although scanning speed had relatively little effect on the line width, the width was affected significantly by the pulse energy.

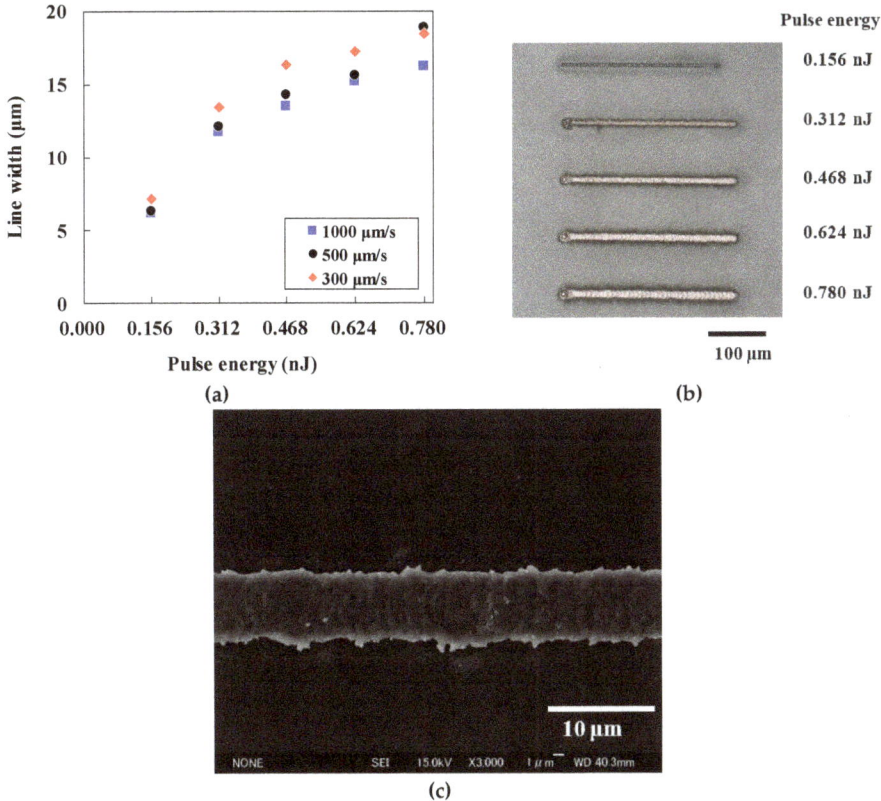

(a)

(b)

(c)

Figure 3. (a) Relationship between line width and laser irradiation conditions, (b) optical microscope image at scanning speed of 1000 μm/s and various pulse energies, and (c) field-emission scanning electron microscopy (FE-SEM) image showing the line width obtained when using a pulse energy of 0.156 nJ and scanning speed of 500 μm/s.

An optical microscope image of the lines for evaluation is shown in Figure 3b. The scanning speed was 1000 μm/s and pulse energy was 0.156−0.780 nJ. We observed line patterns with a copper-like luster.

Figure 3c shows a FE-SEM image of a line pattern fabricated using a pulse energy of 0.156 nJ and a scanning speed of 500 μm/s. We observed the minimum line width of 6.1 μm at a pulse energy of 0.156 nJ and scanning speeds of 500 and 1000 μm/s; this is 7.4 times wider than the focal spot diameter. The greater line width appears to be due to the diffusion of thermal energy around the irradiated region. However, the line width is smaller than that previously obtained using CO_2 laser reduction of a GACu complex [9] and using laser direct writing in air [1,5,7,8].

The direct writing of finer Cu wires is advantageous for the fabrication of integrated microdevices and of connections between electrodes. We expect that the line width can be reduced further by employing tightly focused femtosecond laser pulses using a high-NA objective lens.

3.3. Resistivities of the Line Patterns

The resistivity of the line patterns was obtained by measuring the resistances and cross sections of the line patterns formed to connect Cu thin-film pads on a glass substrate. The size of each pad was 2 mm × 2 mm, and the gap between them was 110 μm which was the length of the line. The film thickness was ~300 μm. The resistance was less than 1 mΩ. Figure 4a shows a typical line pattern connecting the two Cu-thin-film pads, which were fabricated using lithography and sputtering methods. Figure 4b shows the resistivity as a function of pulse energy at various scanning speeds. Compared with the resistivities of the lines, the resistance of the Cu thin-film pads (<1 mΩ) is negligible. The minimum resistivity was 2.43×10^{-6} Ω·m for a line pattern formed with a pulse energy of 0.468 nJ and a scanning speed of 500 μm/s when the line width was ~14 μm as shown in Figure 3a. The line thickness was also estimated to be ~600 nm from the cross-sectional profile shown in Figure 4c. The center of the line was thinner than the sides. This indicates that the center was well sintered because of the higher central intensities of the laser pulses. This line width is significantly smaller than that obtained in the previous work, i.e., 200 μm [9]. However, the resistivity of the line pattern fabricated using femtosecond laser reduction was larger than the resistivity obtained in previous reports [9]. The resistivity increased at higher and lower pulse energies. The femtosecond laser pulse-induced rapid heating produced a combination of phenomena, such as the balance between reduction and reoxidation of Cu, sintering, and heat accumulation. The thermal history of the irradiated region must therefore be taken into account in order to determine the generated composition such as Cu and copper oxides. As a result, the line patterns are made of various composites of Cu and copper oxides under different laser-irradiation conditions.

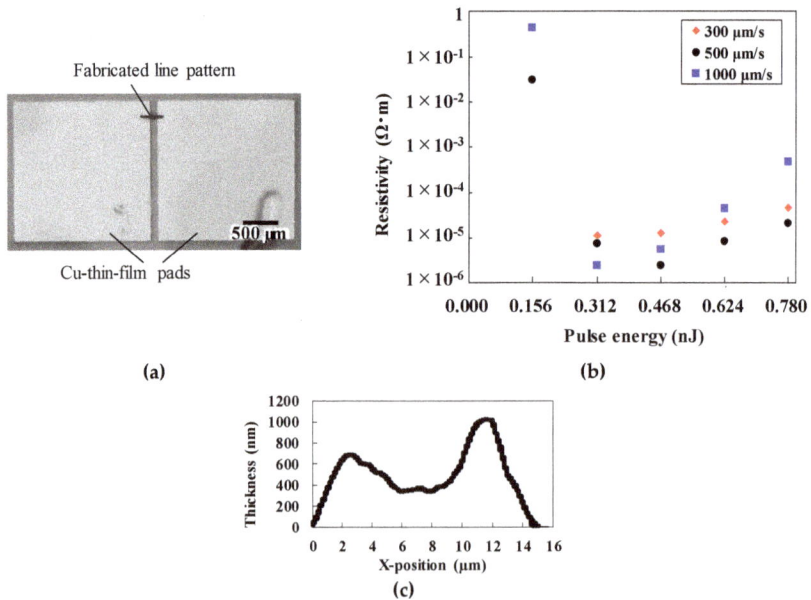

(a)

(b)

(c)

Figure 4. (a) Optical microscope image of a typical line pattern connecting two Cu thin film pads. (b) Resistivity of micropatterns fabricated under various laser irradiation conditions. (c) Cross-sectional profile of a line pattern produced at scanning speed of 500 μm/s and pulse energy of 0.468 nJ.

3.4. Crystal Structures of the Micropatterns

We now discuss the crystal structure of the micropatterns fabricated under various laser-irradiation conditions. The micropatterns measured 600 μm × 900 μm. The raster pitch of the micropattern was determined to be 1 μm by considering the laser focal spot of 1.3 μm. Figure 5a–c shows the XRD spectra of the micropatterns fabricated with scanning speeds of 300, 500, and 1000 μm/s, respectively. All spectra exhibit the diffraction peaks for Cu and Cu_2O.

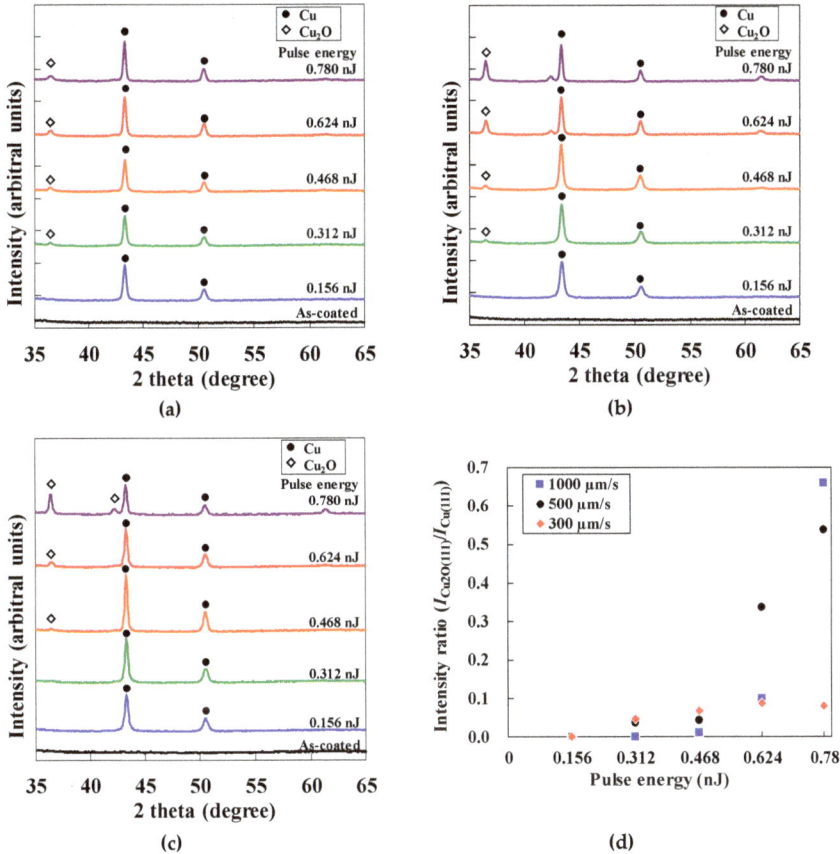

Figure 5. XRD spectra of fabricated micropatterns at a scanning speed of (**a**) 300 μm/s, (**b**) 500 μm/s, and (**c**) 1000 μm/s. (**d**) Intensity ratio of Cu_2O to Cu as a function of pulse energy.

To compare the generation of Cu and Cu_2O under different laser-irradiation conditions, we formed the XRD intensity ratio, for which the peak XRD intensity $I_{Cu2O(111)}$ of $Cu_2O(111)$ was divided by that of Cu(111), $I_{Cu(111)}$ (i.e., $I_{Cu2O(111)}/I_{Cu(111)}$). Figure 5d shows this intensity ratio as a function of pulse energy. The generation of Cu_2O increases with increasing pulse energy for all scanning conditions, which indicates that Cu_2O is generated by re-oxidation of previously generated Cu. The larger amount of Cu_2O generated at high pulse energy is attributed to the grown Cu NPs generated by reduction at low scanning speed and that is difficult to re-oxidize, thereby preventing an increase in Cu_2O.

By accounting for the generation of Cu and Cu_2O, the increase in resistivity at high pulse energy is attributed to re-oxidation of previously generated Cu. In contrast, the increase in resistivity at low pulse energy suggests a lack of reduction of GACu. In general, the use of a short pulse duration prevents re-oxidation [10–12]. Controlling the temperature distribution and history in the line patterns may reduce the resistivity of the line patterns by changing the laser-pulse intensity distribution.

4. Conclusions

Cu-rich micropatterns were fabricated by femtosecond laser pulse-induced reduction of GACu complex.

(1) The minimum line width in the micropatterns was 6.1 μm, which was obtained with a laser-pulse energy of 0.156 nJ and scanning speeds of 500 and 1000 μm/s.

(2) The minimum resistivity of the line pattern was 2.43×10^{-6} Ω·m which was ~10 times greater than that of the pattern formed using a CO_2 laser.

The results of the XRD analysis suggest that the balance of the reduction and the reoxidation of the GACu complex determines the ambient-air generation of highly reduced Cu patterns.

Author Contributions: K.A. and M.M. performed the experiments; K.A., M.M., A.U. and T.O. analyzed the data; M.M. contributed analysis tools; M.M. wrote the paper.

Funding: This study was supported in part by the Nano-Technology Platform Program (Micro-NanoFabrication), the Leading Initiative for Excellent Young Researchers of the Ministry of Education, Culture, Sports, Science and Technology, Japan (MEXT), the 10th "Shiseido Female Researcher Science Grant", and JSPS KAKENHI Grant number JP16H06064.

Conflicts of Interest: The authors declare no conflicts of interest.

References

1.   Kang, B.; Han, S.; Kim, H.J.; Ko, S.; Yang, M. One-step fabrication of copper electrode by laser-induced direct local reduction and agglomeration of copper oxide nanoparticle. *J. Phys. Chem. C* **2011**, *115*, 23664–23670. [CrossRef]

2.   Lee, H.; Yang, M. Effect of solvent and PVP on electrode conductivity in laser-induced reduction process. *Appl. Phys. A* **2015**, *119*, 317–323. [CrossRef]

3.   Lee, D.; Paeng, D.; Park, H.K.; Grigoropoulos, C.P. Vacuum-free, maskless patterning of Ni electrodes by laser reductive sintering of NiO nanoparticle ink and its application to transparent conductors. *ACS Nano* **2014**, *8*, 9807–9814. [CrossRef] [PubMed]

4.   Paeng, D.; Lee, D.; Yeo, J.; Yoo, J.H.; Allen, F.I.; Kim, I.; So, H.; Park, H.K.; Minor, A.M.; Grigoropoulos, C.P. Laser-induced reductive sintering of nickel oxide nanoparticles under ambient conditions. *J. Phys. Chem. C* **2015**, *119*, 6363–6372. [CrossRef]

5.   Mizoshiri, M.; Arakane, S.; Sakurai, J.; Hata, S. Direct writing of Cu-based micro-temperature detectors using femtosecond laser reduction of CuO nanoparticles. *Appl. Phys. Express* **2016**, *9*, 036701. [CrossRef]

6.   Mizoshiri, M.; Ito, Y.; Sakurai, J.; Hata, S. Direct fabrication of Cu/Cu_2O composite micro-temperature sensor using femtosecond laser reduction patterning. *Jpn. J. Appl. Phys.* **2016**, *55*, 06GP05. [CrossRef]

7.   Joo, M.; Lee, B.; Jeong, S.; Lee, M. Comparative studies on thermal and laser sintering for highly conductive Cu films printable on plastic substrate. *Thin Solid Films* **2012**, *520*, 2878–2883. [CrossRef]

8.   Joo, M.; Lee, B.; Jeong, S.; Kim, Y.; Lee, M. Enhanced surface coverage and conductivity of Cu complex ink-coated films by laser sintering. *Thin Solid Films* **2014**, *564*, 264–268.

9.   Ohishi, T.; Kimura, R. Fabrication of copper wire using glyoxylic acid copper complex and laser irradiation in air. *Mater. Sci. Appl.* **2015**, *6*, 799–808. [CrossRef]

10. Qin, G.; Watanabe, A.; Tsukamoto, H.; Yonezawa, T. Copper film prepared from copper fine particle paste by laser sintering at room temperature: Influences of sintering atmosphere on the morphology and resistivity. *Jpn. J. Appl. Phys.* **2014**, *53*, 096501. [CrossRef]

11. Soltani, A.; Vahed, B.K.; Mardoukhi, A.; Mäbtysalo, M. Laser sintering of copper nanoparticles on top of silicon substrates. *Nanotechnology* **2016**, *27*, 035203. [CrossRef] [PubMed]

12. Mizoshiri, M.; Kondo, Y. Direct writing of Cu-based fine micropatterns using femtosecond laser pulse-induced sintering of Cu_2O nanospheres. *Jpn. J. Appl. Phys.* **2019**, *58*, SDDF05. [CrossRef]

micromachines

MDPI

Article

Molecular Dynamics Simulation of the Influence of Nanoscale Structure on Water Wetting and Condensation

Masaki Hiratsuka *, Motoki Emoto, Akihisa Konno and Shinichiro Ito

Department of Mechanical Engineering, Kogakuin University, Tokyo 163-8677, Japan
* Correspondence: hiratsuka@cc.kogakuin.ac.jp; Tel.: +81-42-628-4491

Received: 23 May 2019; Accepted: 29 August 2019; Published: 31 August 2019

Abstract: Recent advances in the microfabrication technology have made it possible to control surface properties at micro- and nanoscale levels. Functional surfaces drastically change wettability and condensation processes that are essential for controlling of heat transfer. However, the direct observation of condensation on micro- and nanostructure surfaces is difficult, and further understanding of the effects of the microstructure on the phase change is required. In this research, the contact angle of droplets with a wall surface and the initial condensation process were analyzed using a molecular dynamics simulation to investigate the impact of nanoscale structures and their adhesion force on condensation. The results demonstrated the dependence of the contact angle of the droplets and condensation dynamics on the wall structure and attractive force of the wall surface. Condensed water droplets were adsorbed into the nanostructures and formed a water film in case of a hydrophilic surface.

Keywords: functional surface; condensation; molecular dynamics; wettability; nanoscale structure

---

1. Introduction

With recent advances in the micro- and nanoscale processing and measurement technology, it has become possible to add fine structures to surfaces [1–3]. These microstructures are known to have significant effects on wettability of liquids [4–6] and are expected to be able to control water–surface interactions and wettability by changing the size of wall structures [7,8]. Wettability of metals and nanostructures changes the friction of objects, chemical reaction on the surface, and crystallization of proteins [9–15]. Mirco-nanosurface is also expected to be used as a highly efficient heat transport device. In case of the condensation heat transfer, micro-nanostructure of a condensation surface is quite essential for achieving a high heat transfer [16]. The condensation growth morphology depends on micro-nanoscale surface topography [17,18]. Also, the condensation form, filmwise and dropwise condensation, is controlled by the surface structures [19,20]. For this reason, the impact of surface structure and wettability on the condensation characteristics has been experimentally investigated [21–29]. However, it is still difficult to observe the initial stage of liquid condensation on nanoscale surfaces directly and analyze the mechanism of the observed phenomena using experimental methods alone. Wettability is affected not only by the shape and size of asperities but also the molecular scale crystal structure of materials [30,31]. Therefore, detailed observations at the atomic scale are required to understand the mechanism of condensation on nanoscale structures.

Analysis using a molecular simulation is one way to elucidate such nanoscale phenomena [32,33]. In previous studies, wettability and the contact angle of droplets with nanoscale surfaces were analyzed using molecular dynamics simulations [6,34,35]. They demonstrated that the Wenzel state [36] and Cassie-Baxter state [37] can be observed depending on the size and spacing of nanostructures, as well as parameters of the molecular interactions between water and surface molecules. The schematic diagrams of the Wenzel state and Cassie-Baxter state are shown in Figure 1. Larger adsorption energy between a wall and water puts the former in the Wenzel state. Also, the smaller the height of a structure, the lower the gap between its wall and droplets, which puts the structure in the Wenzel state [34]. While molecular dynamics calculations have been performed for droplets on nanosurfaces, there are not enough studies focusing on condensation, except only a few investigating condensations on nanostructures under limited conditions [38,39]. These latter studies analyzed the temperature change during condensation and heat flux on surfaces. In the condensation heat transfer on wall surfaces, the size of the structure, material, and water–solid interaction are considered to play an important role. Widely analyzing the size of structures and their interaction with water is essential for understanding the micro- and nanostructure effects on water condensation. Therefore, in this study, we performed a molecular dynamics simulation to reveal the condensation mechanism of water droplets from vapor on nanoscale structures. In addition, we analyzed the contact angles in the static state in relation to the condensation types. The wall–water interaction parameter was changed in the range of hydrophilic to hydrophobic region.

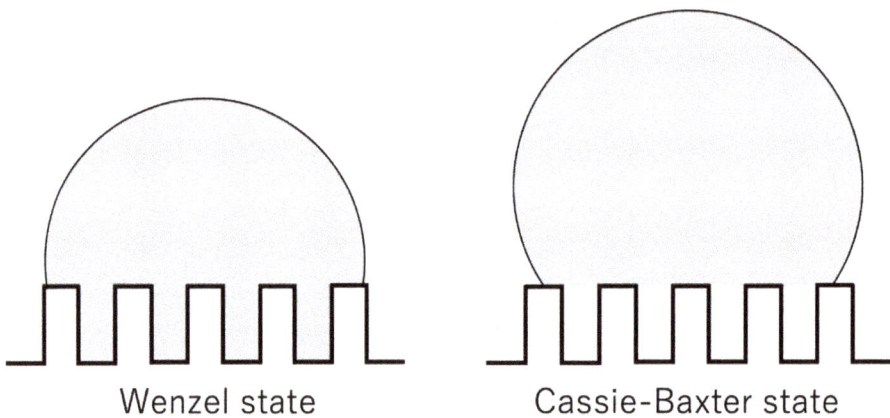

Figure 1. Schematic diagrams of Wenzel state and Cassie-Baxter state.

2. Computational Methods

We performed a molecular dynamics simulation to analyze the effect of the surface structure on the water wettability and condensation process. Coulombic and van der Waals interactions were treated as the intermolecular interactions. The TIP4P model [40], which is the four-site model, was employed as a water molecule model and the Lennard-Jones particle was used as the wall surface. To investigate the effect of adhesion force of wall on the wetting and condensation, the parameter ε was changed as shown in Table 1, ε = 19.74 kJ/mol (hydrophilic) to 0.06168 kJ/mol (hydrophobic). This approach that changes the ε parameter is similar to the previous molecular dynamics simulation in graphene [35]. The molecular size parameter was set to σ = 0.233 nm, the distance at which the intermolecular potential between the two particles is zero. The LJ parameter ε = 19.74 kJ/mol, the upper value of ε, and σ = 0.233 nm are same values used for copper, a typical hydrophilic metal [27]. The cutoff radius for the van der Waals interaction was 1.3 nm, and timestep was set to 2.0 fs. The Ewald method was employed for the calculation of Coulombic interaction. The calculation was performed using GROMACS [41,42]. The simulation was performed under constant number of molecules, volume, and temperature (NVT). The temperature was controlled by Nose–Hoover thermostat [43,44].

Table 1. The parameters of ε in the calculation. The ε = 19.74 kJ/mol is the case of copper.

Ratio	ε (kJ/mol)
1.0	19.74
0.9	17.76
0.8	15.79
0.7	13.82
0.6	11.84
0.5	9.869
0.4	7.895
0.3	5.921
0.2	3.948
0.1	1.974
1/20	0.9869
1/40	0.4935
1/80	0.2467
1/160	0.1234
1/320	0.06168

Three patterns of a nanostructure with different heights and a flat surface were employed as the microfabricated surface as shown in Figure 2. The heights of the nanostructure were set as multiples of the length of the lattice constant of fcc copper 0.362 nm (0.724 nm, 1.448 nm, and 2.896 nm, respectively). The surface wall was a 10 nm square, 1.448 nm in thickness, under periodic boundary condition. The length of the height direction of the simulation box was 50 nm.

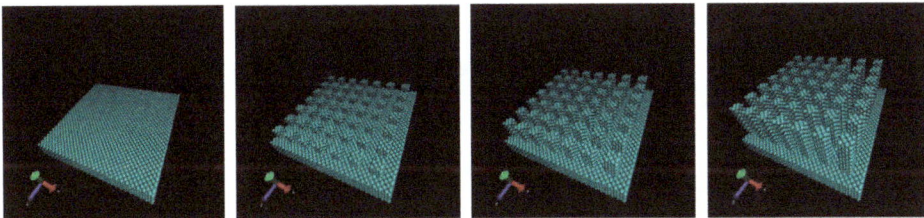

Figure 2. Prepared surface structure. (flat, asperity height 0.724 nm, asperity height 1.448 nm, and asperity height 2.896 nm).

The contact angle of droplets deposited on the wall was measured using the half-angle method [45]. The schematic figure of the determination of contact angle θ on the nanosurface in this study is illustrated as Figure 3. The contact angle was determined as the angle of the line from the triple phase point to the apex of the drop and the line of the top of the wall. In the initial state of the simulation, a droplet was placed 0.5 nm from the wall and its natural adsorption on the wall was analyzed. To understand the effect of the size of the droplet on the wall, two diameters of the droplet, about 5 nm (2259 water molecules) and about 6 nm (3787 water molecules) were explored. The contact angle was calculated after 4 ns to reach the equilibrium at 300 K.

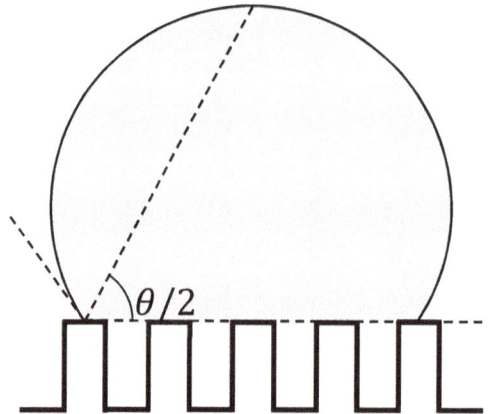

Figure 3. Schematic diagram of contact angle θ measurement on nanostructures by half angle method.

The condensation process of water vapor on the nanostructured surface was calculated by 5 ns simulation under 300 K. The initial structure of the calculation was prepared by the 2 ns calculation under 600 K for 2259 water molecules on the surface as shown in Figure 4. We changed the LJ potential parameter ε in three patterns, 19.74 kJ/mol, 1.974 kJ/mol, 0.06168 kJ/mol. Three calculations were performed for each condition to obtain average values.

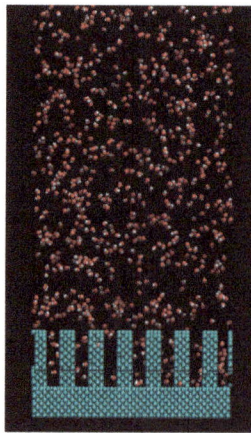

Figure 4. Snapshot of water vapor and nanostructure prepared as the initial condition of the condensation simulation.

3. Results and Discussions

Figure 5 and Tables 2 and 3 show the calculated contact angles when the droplets with sizes of 5 nm and 6 nm were on the wall surface. Both contact angle dependence on the water–wall interaction and height of pillar showed similar trends with the previous study on graphene [35]. In Tables 2 and 3, the numbers on the gray background correspond to the Wenzel state, the numbers without any background represent the Cassie-Baxter state, while no number means no droplet. The contact angle was determined as the angle at the top of the surface of asperity. The part where the angle could not be calculated corresponds to the area where the solid–liquid part was not formed stably because the liquid spread over the entire surface. Figure 6 shows a snapshot of the case, where water molecules spread into a film and the contact angle could not be determined on the plane of $\varepsilon = 19.74$ kJ/mol. When the adsorption force of the wall was large, water spread into a film on both the flat and uneven surfaces. Even when there was unevenness, water adhered to the available contact area and did not form droplets. The process of forming such a liquid film is consistent with previous molecular simulations on copper surfaces [39]. This behavior is similar to that of macroscopic films spread thinly when droplets are adsorbed on a hydrophilic surface [46,47]. Two layers of water molecules were found on the surface in the case of a film formed by 5 nm droplets on the flat surface. The hydrophilic surface adsorbed water molecules and aligned them. The boundary ε for the wetting state changing from Wenzel to Cassie-Baxter was 1.974 or 0.9869 kJ/mol, depending on the pillar height. There was no difference in the size of the droplets.

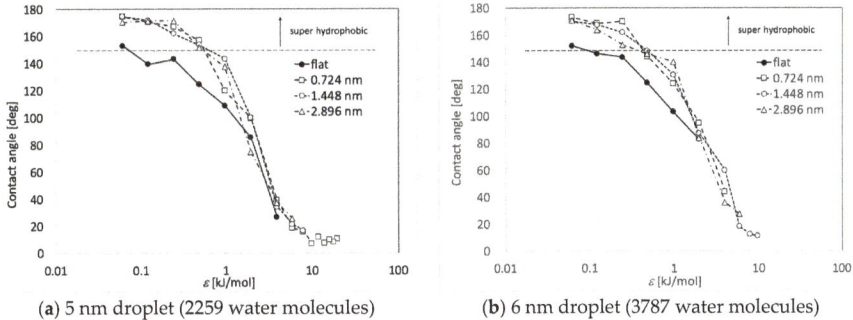

(**a**) 5 nm droplet (2259 water molecules)   (**b**) 6 nm droplet (3787 water molecules)

Figure 5. Relation between the water–wall interaction and the contact angle. The surface with asperity resulted in the increase of the contact angle.

Table 2. Contact angle and wetting state of 5-nm droplets on each surface. The numbers on the gray background correspond to the Wenzel state, whereas the numbers without background correspond to the Cassie state.

ε (kJ/mol)	Flat	0.724 nm	1.448 nm	2.896 nm
19.74	-	10.6	-	-
17.76	-	8.1	-	-
15.79	-	10.0	-	-
13.82	-	7.5	-	-
11.84	-	11.8	-	-
9.869	-	7.0	-	-
7.895	-	15.8	16.7	-
5.921	-	18.5	21.1	25.3
3.948	26.5	39.0	34.3	37.2
1.974	85.4	99.7	99.9	74.4
0.9869	108.8	120.0	143.4	137.5
0.4935	124.4	157.1	153.7	151.8
0.2467	143.3	167.0	162.0	171.3
0.1234	139.6	170.5	171.9	171.4
0.06168	153.0	174.5	174.6	170.5

Table 3. Contact angle and wetting state of 6-nm droplets on each surface. The numbers on the gray background correspond to the Wenzel state, whereas the numbers without background correspond to the Cassie state.

ε (kJ/mol)	Flat	0.724 nm	1.448 nm	2.896 nm
19.74	-	-	-	-
17.76	-	-	-	-
15.79	-	-	-	-
13.82	-	-	-	-
11.84	-	-	-	-
9.869	-	-	10.9	-
7.895	-	-	12.3	-
5.921	-	-	18.1	27.4
3.948	-	44.0	59.6	36.0
1.974	82.6	95.0	87.3	83.8
0.9869	103.2	124.1	130.6	140.3
0.4935	124.6	144.1	148.2	145.9
0.2467	143.7	170.1	161.8	153.0
0.1234	146.3	168.5	167.5	163.7
0.06168	152.1	173.2	170.3	170.7

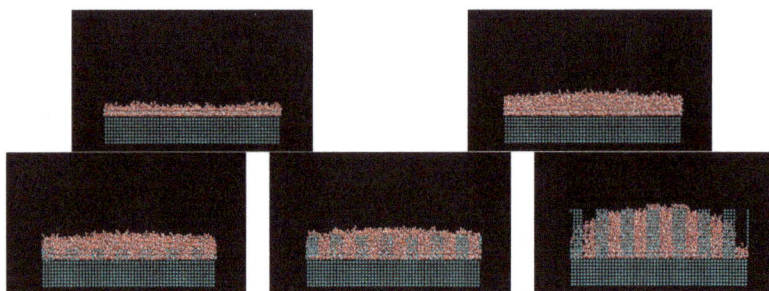

Figure 6. Snapshot of the water film on the flat and nanostructured surfaces with the Lennard-Jones parameter ε = 19.74 kJ/mol (the left upper: 2259 water molecules, others: 3787 water molecules).

Figure 7 shows the change in the wetting state when asperity changed for a wall adsorption force of ε = 3.948 kJ/mol and a droplet with a diameter of 6 nm. While the liquid spreads over the whole plane on a flat wall, the Wenzel state is manifested by adding unevenness. Similarly, Figure 8 shows how the Wenzel and Cassie states are switched depending on the size of unevenness for the wall adsorption energy ε = 0.9869 kJ/mol. Overall, the contact angles were increased by the nanostructures by about 10° to 40°. The contact angle tended to increase with asperity and as the interaction between water molecules and the wall surface decreased. It was also possible to estimate how the liquid film, Wenzel state, and Cassie state changed depending on the surface adsorption force and nanoscale unevenness size when droplets adhered to the solid surface.

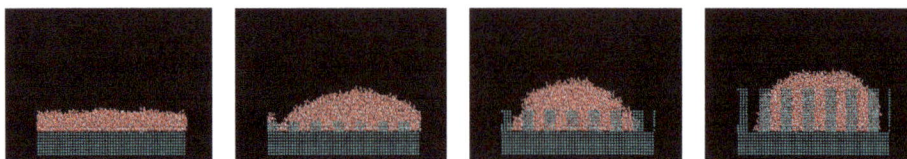

Figure 7. Droplets on nanostructures with ε = 3.948 kJ/mol.

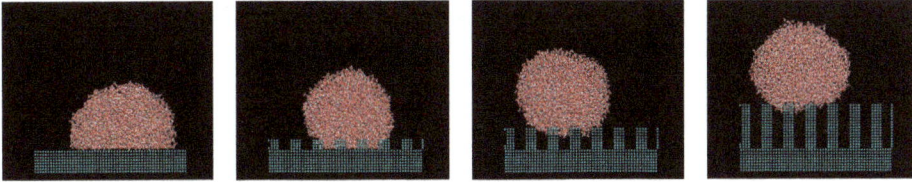

Figure 8. Droplets on nanostructures with $\varepsilon = 0.9869$ kJ/mol.

Figures 9–11 show the appearance of the convex surface in the case of $\varepsilon = 19.74$ kJ/mol, 1.974 kJ/mol, and 0.06168 kJ/mol. These parameters correspond to the liquid film, Wenzel state, and Cassie-Baxter state, respectively, in the calculation of the water droplets. When water molecules cooled down, both condensation near the surface and in the vapor could be observed. Figure 9 shows the snapshot of condensation with a strong water–surface interaction. Condensed water molecules formed a liquid film and uniformly attached to the inner wall of asperities. The surface of the hydrophilic nanostructure was wet in the initial stage of the condensation process. In addition, the small water droplets formed in the water vapor were observed to be absorbed into the asperity surface. Figure 12 shows the absorption behavior of water droplets intruding into the inside of asperities. It was found that the time scale of droplet adsorption is several tens to hundreds of ps. In such a hydrophilic nanostructure, water molecules spreading thinly inside asperities formed an orderly structure different from a bulk liquid. Figure 13 illustrates a snapshot of the two-dimensional structure of water observed in the nanostructured surface. This type of ordered structure is unique to confined systems such as inside nanotubes and graphene plates [48,49]. These results indicate that water molecules in nanostructured hydrophilic metal surfaces form unusual phase structures similar to other confined systems. Figure 10 demonstrates condensation of water on a wall with a low interaction level. When the interaction level became smaller, smaller droplets gradually attached to the solid surface but were not uniformly spread. Several droplets formed and gradually integrated. Figure 11 demonstrates the case of condensation on a hydrophobic surface with a very small interaction level. Even when a small number of water molecules formed a few clusters within the asperity, they gradually discharged to the outside of the asperity. In the end, almost no water molecules were left inside the asperities, and the droplets were attached to the surface.

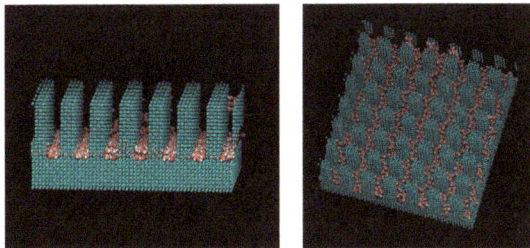

Figure 9. Snapshot of water molecules during the condensation on the $\varepsilon = 19.74$. kJ/mol surface from the different viewpoints. The water molecules are adsorbed into the pillar and formed water film.

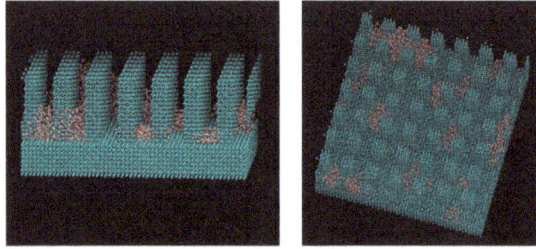

Figure 10. Snapshot of water molecules during the condensation on the $\varepsilon = 1.974$. kJ/mol surface from the different viewpoints. The water molecules formed small clusters in the pillar.

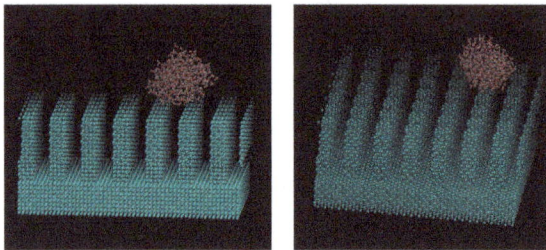

Figure 11. Snapshot of water molecules during the condensation on the $\varepsilon = 0.06168$. kJ/mol surface from the different viewpoints. The water droplet did not enter the nanostructure.

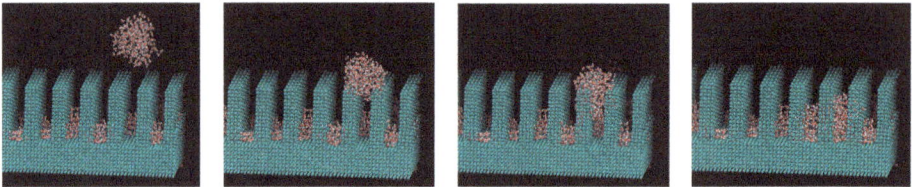

Figure 12. Absorption behavior of water droplets intruding into the inside of asperities with $\varepsilon = 1.976$ kJ/mol. (0 s, 35 ps, 90 ps, 350 ps).

Figure 13. Snapshot of the two-dimensional structure of water observed in a nanostructured surface with $\varepsilon = 19.74$. kJ/mol.

Figure 14 shows the average number of water molecules inside the surface nanostructure for each case. The number of water molecules was increased for the cases of $\varepsilon = 19.74$ kJ/mol and $\varepsilon = 1.974$ kJ/mol by adsorption on the surface. For the first 1 ns, isolated water molecules near the surface adsorbed on the nanostructures continuously. After 1 ns, the increase in the number of water molecules showed jumps due to the adsorption of water droplet formed in the vapor phase. On the other hand, the number of water molecules was decreased in the case of $\varepsilon = 0.06168$ kJ/mol. After 1 ns, a small number of water molecules was trapped in the nanostructure. Figure 15 shows the mean square displacement (MSD) of the water molecules in the nanostructure. The MSD of the case of $\varepsilon = 19.74$ kJ/mol and $\varepsilon = 1.974$ kJ/mol is small because the water molecules on the surface were almost fixed or restricted in the droplet. The MSD for the case of $\varepsilon = 0.06168$ kJ/mol is much larger than the others. The small number of water molecules trapped in the nanostructure moved quickly on the surface.

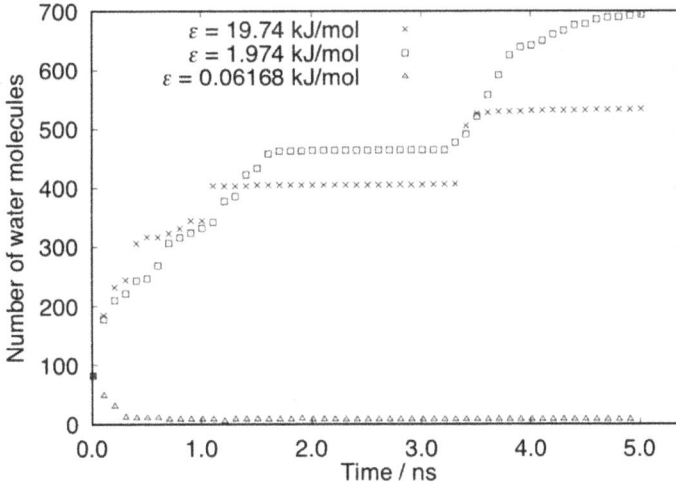

Figure 14. Number of water molecules in the nanostructure on surfaces.

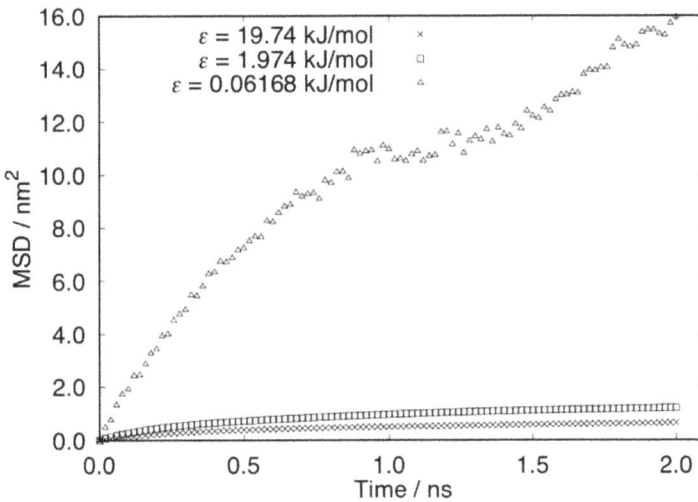

Figure 15. Mean square displacement (MSD) of water molecules in the nanostructure on surfaces.

As demonstrated here, the three types of condensation behavior—film, droplet, and discharge—appeared according to the difference in the strength of the surface interaction. In particular, when the adsorptive force was large, as is the case with copper, water molecules aggregated on the surface of the asperity and prepared a water film. The differences in the condensation behavior affected the condensation speed and diffusion of water molecules on the surface.

4. Conclusions

The impact of the wall nanostructure and adsorption force on the contact angle of droplets and condensation was analyzed using molecular dynamics simulation. As a result, the dependence of the contact angle and condensation behavior on the microfabrication shape and size of the wall was revealed. As the condensation behavior, the liquid film formation, droplet adsorption in the structure, and droplet discharge process were observed. The water molecules adsorbed on the surfaces showed little diffusion in the case of $\varepsilon = 19.74$ kJ/mol and $\varepsilon = 1.974$ kJ/mol. In addition, a two-dimensional structure of water molecules spread into the fine structure was observed.

Author Contributions: Conceptualization, M.H.; Investigation, M.H. and M.E.; Supervision, A.K. and S.I.; Writing original draft, M.H.

Funding: This research received no external funding.

Conflicts of Interest: The authors declare no conflict of interest.

References

1. Biró, L.P.; Nemes-Incze, P.; Lambin, P. Graphene: Nanoscale processing and recent applications. *Nanoscale* **2011**, *4*, 1824–1839.
2. Tan, X.; Tao, Z.; Yu, M.; Wu, H.; Li, H. Anti-Reflectance Optimization of Secondary Nanostructured Black Silicon Grown on Micro-Structured Arrays. *Micromachines* **2018**, *9*, 385. [CrossRef] [PubMed]
3. Meng, J.; Dong, X.; Zhao, Y.; Xu, R.; Bai, X.; Zhou, H. Fabrication of a Low Adhesive Superhydrophobic Surface on Ti_6Al_4V Alloys Using TiO_2/Ni Composite Electrodeposition. *Micromachines* **2019**, *10*, 121. [CrossRef] [PubMed]
4. Yoshimitsu, Z.; Nakajima, A.; Watanabe, T.; Hashimoto, K. Effects of Surface Structure on the Hydrophobicity and Sliding Behavior of Water Droplets. *Langmuir* **2002**, *18*, 5818–5822. [CrossRef]
5. Wang, J.; Liu, M.; Ma, R.; Wang, Q.; Jiang, L. In Situ Wetting State Transition on Micro- and Nanostructured Surfaces at High Temperature. *ACS Appl. Mater. Interfaces* **2014**, *6*, 15198–15208. [CrossRef]
6. Chen, S.; Wang, J.; Chen, D. States of a Water Droplet on Nanostructured Surfaces. *J. Phys. Chem. C* **2014**, *118*, 18529–18536. [CrossRef]
7. Yong, X.; Zhang, L.T. Nanoscale Wetting on Groove-Patterned Surfaces. *Langmuir* **2009**, *25*, 5045–5053. [CrossRef]
8. Introzzi, L.; Fuentes-Alventosa, J.M.; Cozzolino, C.A.; Trabattoni, S.; Tavazzi, S.; Bianchi, C.L.; Schiraldi, A.; Piergiovanni, L.; Farris, S. "Wetting Enhancer" Pullulan Coating for Antifog Packaging Applications. *ACS Appl. Mater. Interfaces* **2012**, *4*, 3692–3700. [CrossRef]
9. Ho, T.A.; Papavassiliou, D.V.; Lee, L.L.; Striolo, A. Liquid water can slip on a hydrophilic surface. *Proc. Natl. Acad. Sci. USA* **2011**, *108*, 16170–16175. [CrossRef]
10. Wang, C.; Wen, B.; Tu, Y.; Wan, R.; Fang, H. Friction Reduction at a Superhydrophilic Surface: Role of Ordered Water. *J. Phys. Chem. C* **2015**, *119*, 11679–11684. [CrossRef]
11. Tocci, G.; Joly, L.; Michaelides, A. Friction of Water on Graphene and Hexagonal Boron Nitride from ab initio Methods: Very Different Slippage Despite Very Similar Interface Structures. *Nano Lett.* **2015**, *14*, 6872–6877. [CrossRef] [PubMed]
12. Lu, P.; Liu, X.; Zhang, C. Electroosmotic Flow in a Rough Nanochannel with Surface Roughness Characterized by Fractal Cantor. *Micromachines* **2017**, *8*, 190. [CrossRef]
13. Gao, W.; Zhang, X.; Han, X.; Shen, C. Role of Solid Wall Properties in the Interface Slip of Liquid in Nanochannels. *Micromachines* **2018**, *9*, 663. [CrossRef] [PubMed]
14. Yang, Z.; Shi, B.; Lu, H.; Xiu, P.; Zhou, R. Dewetting Transitions in the Self-Assembly of Two Amyloidogenic β-Sheets and the Importance of Matching Surfaces. *J. Phys. Chem. B* **2011**, *115*, 11137–11144. [CrossRef] [PubMed]

15. Weibel, D.E.; Michels, A.F.; Feil, A.F.; Amaral, L.; Teixeira, S.R.; Horowitz, F. Adjustable Hydrophobicity of Al Substrates by Chemical Surface Functionalization of Nano/Microstructures. *J. Phys. Chem. C* **2010**, *114*, 13219–13225. [CrossRef]

16. Miljkovic, N.; Wang, E.N. Condensation heat transfer on superhydrophobic surfaces. *MRS Bull.* **2013**, *38*, 397–406. [CrossRef]

17. Miljkovic, N.; Enright, R.; Wang, E.N. Effect of Droplet Morphology on Growth Dynamics and Heat Transfer during Condensation on Superhydrophobic Nanostructured Surfaces. *ACS Nano* **2012**, *6*, 1776–1785. [CrossRef]

18. Enright, R.; Miljkovic, N.; Al-Obeidi, A.; Thompson, C.V.; Wang, E. Condensation on Superhydrophobic Surfaces: The Role of Local Energy Barriers and Structure Length Scale. *Langmuir* **2012**, *28*, 14424–14432. [CrossRef]

19. Miljkovic, N.; Enright, R.; Nam, Y.; Lopez, K.; Dou, N.; Sack, J.; Wang, E.N. Jumping-Droplet-Enhanced Condensation on Scalable Superhydrophobic Nanostructured Surfaces. *Nano Lett.* **2013**, *13*, 179–187. [CrossRef]

20. Hou, Y.; Yu, M.; Chen, X.; Wang, Z.; Yao, S. Recurrent Filmwise and Dropwise Condensation on a Beetle Mimetic Surface. *ACS Nano* **2015**, *9*, 71–81. [CrossRef]

21. Erb, R.A. Wettability of Metals under Continuous Condensing Conditions. *J. Phys. Chem.* **1965**, *69*, 1306–1309. [CrossRef]

22. Varanasi, K.K.; Hsu, M.; Bhate, N.; Yang, W.; Deng, T. Spatial control in the heterogeneous nucleation of water. *Appl. Phys. Lett.* **2009**, *95*, 094101.

23. Boreyko, J.B.; Chen, C.H. Self-Propelled Dropwise Condensate on Superhydrophobic Surfaces. *Phys. Rev. Lett.* **2009**, *103*, 184501. [CrossRef] [PubMed]

24. Dietz, C.; Rykaczewski, K.; Fedorov, A.G.; Joshi, Y. Visualization of droplet departure on a superhydrophobic surface and implications to heat transfer enhancement during dropwise condensation. *Appl. Phys. Lett.* **2010**, *978*, 033104.

25. Chen, X.; Wu, J.; Ma, R.; Hua, M.; Koratkar, N.; Yao, S.; Wang, Z. Nanograssed Micropyramidal Architectures for Continuous Dropwise Condensation. *Adv. Funct. Mater.* **2011**, *21*, 4617–4623. [CrossRef]

26. Cheng, J.; Vandadi, A.; Chen, C.L. Condensation heat transfer on two-tier superhydrophobic surfaces. *Appl. Phys. Lett.* **2012**, *101*, 131909. [CrossRef]

27. Lee, S.; Cheng, K.; Palmre, V.; Bhuiya, M.D.M.H.; Kim, K.J.; Zhang, B.J.; Yoon, H. Heat transfer measurement during dropwise condensation using micro/nano-scale porous surface. *J. Heat Mass Transf.* **2013**, *65*, 619–626. [CrossRef]

28. Quang, T.S.B.; Leong, F.Y.; An, H.; Tan, B.H.; Ohl, C.D. Growth and wetting of water droplet condensed between micron-sized particles and substrate. *Sci. Rep.* **2016**, *6*, 30989. [CrossRef] [PubMed]

29. He, M.; Ding, Y.; Chen, J.; Song, Y. Spontaneous Uphill Movement and Self-Removal of Condensates on Hierarchical Tower-Like Arrays. *ACS Nano* **2016**, *6*, 9456–9462. [CrossRef]

30. Zhu, C.; Li, H.; Huang, Y.; Zeng, X.C.; Meng, S. Microscopic Insight into Surface Wetting: Relations between Interfacial Water Structure and the Underlying Lattice Constant. *Phys. Rev. Lett.* **2013**, *110*, 126101. [CrossRef]

31. Xu, Z.; Gao, Y.; Wang, C.; Fang, H. Nanoscale Hydrophilicity on Metal Surfaces at Room Temperature: Coupling Lattice Constants and Crystal Faces. *J. Phys. Chem. C* **2015**, *119*, 20409–20415. [CrossRef]

32. Carrasco, J.; Hodgson, A.; Michaelides, A. A molecular perspective of water at metal interfaces. *Nat. Matt* **2012**, *11*, 667.

33. Van Vreumingen, D.; Tewari, S.; Verbeek, F.; van Ruitenbeek, J.M. Towards Controlled Single-Molecule Manipulation Using "Real-Time" Molecular Dynamics Simulation: A GPU Implementation. *Micromachines* **2018**, *9*, 270. [CrossRef] [PubMed]

34. Koishi, T.; Yasuoka, K.; Fujikawa, S.; Ebisuzaki, T.; Cheng, X. Coexistence and transition between Cassie and Wenzel state on pillared hydrophobic surface. *Proc. Natl. Acad. Sci. USA* **2009**, *206*, 8435–8440. [CrossRef] [PubMed]

35. Koishi, T.; Yasuoka, K.; Zeng, X.C. Molecular dynamics simulation of water nanodroplet bounce back from flat and nanopillared surface. *Langmuir* **2017**, *33*, 10184–10192. [CrossRef]

36. Wenzel, R.N. Resistance of Solid Surfaces to Wetting by Water. *Ind. Eng. Chem.* **1936**, *28*, 988–994. [CrossRef]

37. Cassie, A.B.D.; Baxter, S. Wettability of Porous Surfaces. *Trans. Faraday Soc.* **1944**, *40*, 546–551. [CrossRef]

38. Niu, D.; Tang, G.H. The effect of surface wettability on water vapor condensation in nanoscale. *Sci. Rep.* **2016**, *67*, 19192. [CrossRef]
39. Gao, S.; Liao, Q.; Liu, W.; Liu, Z. Effects of solid fraction on droplet wetting and vapor condensation: A molecular dynamic simulation study. *Langmuir* **2017**, *33*, 12379–12388. [CrossRef]
40. Jorgensen, W.L.; Chandrasekhar, J.; Madura, J.D.; Impey, R.W.; Klein, M.L. Comparison of simple potential functions for simulating liquid water. *J. Chem. Phys.* **1983**, *79*, 926–935. [CrossRef]
41. Spoel, D.V.D.; Lindahl, E.; Hess, B.; Groenhof, G.; Mark, A.E.; Berendsen, H.J.C. GROMACS: Fast, flexible, and free. *J. Comp. Chem.* **2005**, *26*, 1701–1718. [CrossRef] [PubMed]
42. Hess, B.; Kutzner, C.; van der Spoel, D.; Lindahl, E. GROMACS 4: Algorithms for Highly Efficient, Load-Balanced, and Scalable Molecular Simulation. *J. Chem. Theory Comput.* **2008**, *4*, 435–447. [CrossRef] [PubMed]
43. Nosé, S. A unified formulation of the constant temperature molecular-dynamics methods. *J. Chem. Phys.* **1984**, *81*, 511–519. [CrossRef]
44. Hoover, W.G. Canonical dynamics: Equilibrium phase-space distributions. *Phys. Rev. A* **1985**, *31*, 1695–1697. [CrossRef] [PubMed]
45. Gu, H.; Wang, C.; Gong, S.; Mei, Y.; Li, H.; Ma, W. Investigation on contact angle measurement methods and wettability transition of porous surfaces. *Surf. Coat. Technol.* **2016**, *292*, 72–77. [CrossRef]
46. Wang, C.; Lu, H.; Wang, Z.; Xiu, P.; Zhou, B.; Zuo, G.; Wan, R.; Hu, J.; Fang, H. Stable Liquid Water Droplet on a Water Monolayer Formed at Room Temperature on Ionic Model Substrates. *Phys. Rev. Lett.* **2009**, *103*, 137801. [CrossRef]
47. Yuan, Q.; Zhao, Y.P. Precursor Film in Dynamic Wetting, Electrowetting, and Electro-Elasto-Capillarity. *Phys. Rev. Lett.* **2010**, *104*, 246101. [CrossRef] [PubMed]
48. Zhu, Y.; Wang, F.; Bai, J.; Xeng, X.C.; Wu, H. Compression Limit of Two-Dimensional Water Constrained in Graphene Nanocapillaries. *ACS Nano* **2015**, *9*, 12197–12204. [CrossRef]
49. Gao, Z.; Giovambattista, N.; Sahin, O. Phase Diagram of Water Confined by Graphene. *Sci. Rep.* **2018**, *8*, 6228. [CrossRef]

micromachines

MDPI

Article

4D Printing of Multi-Hydrogels Using Direct Ink Writing in a Supporting Viscous Liquid

Takuya Uchida and Hiroaki Onoe *

School of Integrated Design Engineering, Graduate School of Science and Technology, Keio University, 3-14-1 Hiyoshi, Kouhoku-ku, Yokohama, Kanagawa 223-8522, Japan
* Correspondence: onoe@mech.keio.ac.jp; Tel.: +81-45-566-1507

Received: 30 May 2019; Accepted: 28 June 2019; Published: 30 June 2019

Abstract: We propose a method to print four-dimensional (4D) stimuli-responsive hydrogel structures with internal gaps. Our 4D structures are fabricated by printing an N-isopropylacrylamide-based stimuli-responsive pre-gel solution (NIPAM-based ink) and an acrylamide-based non-responsive pre-gel solution (AAM-based ink) in a supporting viscous liquid (carboxymethyl cellulose solution) and by polymerizing the printed structures using ultraviolet (UV) light irradiation. First, the printed ink position and width were investigated by varying various parameters. The position of the printed ink changed according to physical characteristics of the ink and supporting liquid and printing conditions including the flow rates of the ink and the nozzle diameter, position, and speed. The width of the printed ink was mainly influenced by the ink flow rate and the nozzle speed. Next, we confirmed the polymerization of the printed ink in the supporting viscous liquid, as well as its responsivity to thermal stimulation. The degree of polymerization became smaller, as the interval time was longer after printing. The polymerized ink shrunk or swelled repeatedly according to thermal stimulation. In addition, printing multi-hydrogels was demonstrated by using a nozzle attached to a Y shape connector, and the responsivity of the multi-hydrogels to thermal-stimulation was investigated. The pattern of the multi-hydrogels structure and its responsivity to thermal-stimulation were controlled by the flow ratio of the inks. Finally, various 4D structures including a rounded pattern, a spiral shape pattern, a cross point, and a multi-hydrogel pattern were fabricated, and their deformations in response to the stimuli were demonstrated.

Keywords: 4D printing; 3D printing; stimuli-responsive hydrogel

1. Introduction

Accompanying the recent advances in three-dimensional (3D) printing technologies, not only static objects but also shape-changing structures have been fabricated by 3D printers using stimuli-responsive materials. These printed structures have been defined as four-dimensional (4D) printed structures by adding one dimension (time variation) to 3D printed structures [1–5]. 4D printed structures can change their shapes and functionalities in response to external stimuli such as light, heat, and pH changes. Thanks to these characteristics, 4D printed objects and machines can be expected to achieve self-assembly [6], self-adaptability [7], and self-repair [8].

In terms of the materials used for 4D printing, stimuli-responsive polymers [9] and hydrogels [10] have mainly been adopted. In particular, stimuli-responsive hydrogels have been applied to drug delivery systems [11] and soft actuators [12,13], owing to their biocompatibility and softness. Using stimuli-responsive hydrogels, 4D microstructures have usually been printed using photolithography [14–18] and deposition printing on substrates [19–23]. For all these methods, fabricated hydrogel structures are mainly layered structures, creating bending or twisting motions for the printed structures. However, it is difficult to fabricate 4D structures with internal gaps or suspended beam structures, both of which can be critical to achieving complex motions and encapsulating materials or micro-channels inside structures for soft robots or medical tools.

Here, we propose a new fabrication method for 4D printing that can fabricate 3D multi-hydrogels structures with internal gaps (Figure 1). We introduce a viscous liquid—carboxymethyl cellulose aq (CMC aq)—as a supporting viscous liquid during printing. As a printing ink, we chose a mixture of a poly-*N*-isopropylacrylamide (pNIPAM) solution, which exhibits a stimuli-responsive (thermo-responsive) shrinking/swelling characteristic after gelation, and a polyacrylamide (pAAM) solution, which does not respond to stimulation. To adjust the viscosity of the pNIPAM and pAAM print ink solution, we added a sodium alginate solution (NaAlg) to the printing ink. The print ink was printed directly through a nozzle in CMC aq (supporting viscous liquid) such that the printed ink can be maintained in the printed position to create 3D patterns with internal gaps. Then, the printed 3D ink patterns can be polymerized using ultraviolet (UV) irradiation. We printed a straight line of ink and investigated the position and width of the printed ink under various conditions. Next, we printed a corner of the ink and investigated the printing resolution. We polymerized the printed ink in the supporting viscous liquid and investigated the degree of gelation and the responsivity to external stimuli. In addition, we polymerized multi-hydrogels and investigated their printed pattern and responsivity to stimuli. Finally, we printed various 4D structures and investigated their responsivity to thermal stimuli. Our method can provide an effective tool for fabricating hydrogel 4D structures with various types of physical or chemical stimuli for applications in soft actuators/robotics and self-assembly/adaptive systems.

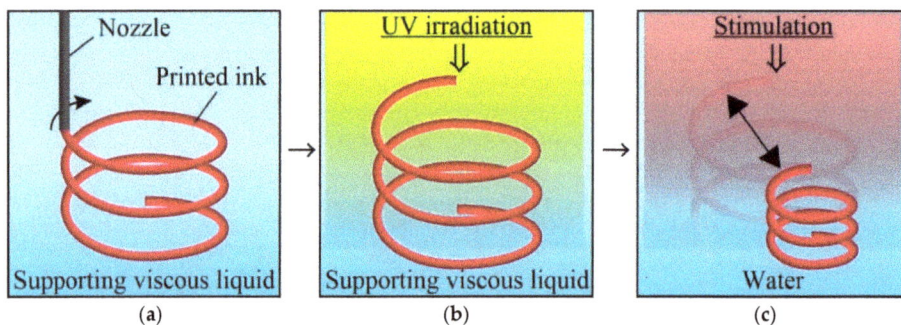

Figure 1. Concept of our proposed method for fabricating 4D structure with internal gaps. (**a**) Pre-gel monomer ink is ejected into supporting viscous liquid to print 3D ink pattern. (**b**) The printed ink is exposed to UV and is polymerized to a obtain 3D hydrogel structure. (**c**) After replaced the supporting liquid into water, the polymerized 3D hydrogel structure deforms in response to stimulation.

2. Materials and Methods

2.1. Materials

N-isopropylacrylamide (NIPAM) (monomer) (113.16 g/mol, 095-03692) and *N,N′*-methylene-bis-acrylamide (BIS) (cross-linking agent) (154.17 g/mol, 134-02352) were purchased from FUJIFILM Wako Chemicals USA, Corp. (Richmond, VA, USA). IRGACURE1173 (photo polymerization initiator) was purchased from BadischAnilin and Soda-Fabrik (Ludwigshafen, Germany). In addition, sodium alginate (NaAlg) (80–120 cp, 194-13321) and acrylamide (AAM) (71.08 g/mol, 019-08011) were purchased from FUJIFILM Wako Chemicals USA, Corp. (Richmond, VA, USA), and carboxymethyl cellulose (CMC) (1000–50000 Pa·s, CMF-150) was purchased from AS ONE Corporation (Osaka, Japan). New Coccine (coloring dye) was purchased from Kyoritsusyokuhin Inc. (Osaka, Japan), and acryloxyethyl thiocarbamoyl rhodamine B (652.2 g/mol, 25404-100) was purchased from Polyscience, Inc. (Warrington, PA, USA). Fluorescence beads (1934417A, 1927586) were purchased from Thermo Fisher Scientific

(Waltham, MA, USA). All chemicals were utilized with no further purification. Deionized water was obtained from a Millipore purification system. Table 1 shows a glossary of the abbreviation of materials.

Table 1. Glossary of abbreviation of materials.

Material	Abbreviation
N-isopropylacrylamide	NIPAM
carboxymethyl cellulose	CMC
Acrylamide	AAM
sodium alginate	NaAlg

2.2. Set Up for 4D Printing

The ink was injected using a syringe pump (LEGATO 180, KD Scientific, Holliston, MA, USA) through a nozzle composed of SUS304 (NN-2225R, TERUMO, Tokyo, Japan) in CMC aq, using *xyz* stages (OSMS20-(XY), OSMS26-(Z), SIGMAKOKI, Tokyo, Japan). The nozzle was fixed with a jig. The program of the stages was set using sample103 (SIGMAKOKI). Figure 2 illustrates the setup for our printing system.

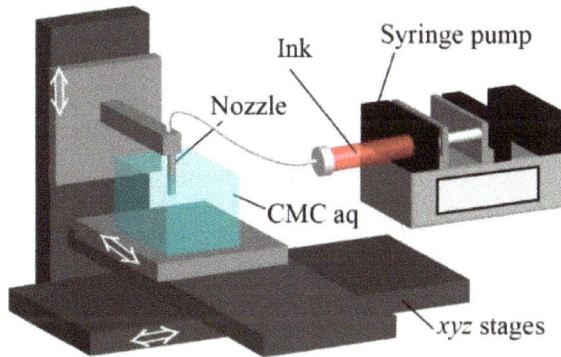

Figure 2. Schematic of setup. A nozzle is fixed on a z stage, and a container of carboxymethyl cellulose aq (CMC aq) is fixed on *xy* stages. By moving the *xyz* stages, the pre-gel ink is ejected via the nozzle into CMC aq by a syringe pump.

2.3. Evaluation of Printed Ink Patterns in Supporting Material

The printing ink for our 4D printing was composed of 10% (*w/w*) NIPAM monomer, 0.01% (*w/w*) BIS, 1% (*w/w*) New Coccine, 0.5% (*w/w*) IRGACURE1173, and 1–3% (*w/w*) NaAlg. The supporting viscous liquid, 0.4–1.6% (*w/w*) CMC was tested (concentration of CMC: C_{CMC}). To investigate printing performance, we printed a straight line (20 mm in length) of printing ink under various conditions, as follows (Table 2). The flow rate of the printing ink, Q, was 0.5–1.5 µL/s, and the stage speed, v, was 0.5–1.5 mm/s. In addition, the diameter and depth of the nozzle, d and h, were 400–800 µm and 5–10 mm, respectively. Table 2 shows the values of the all parameters.

Table 2. Value of the all parameters.

Parameter	Value	Unit
Concentration of NaAlg, C_{NaA}	1.0, 2.0, 3.0	% (*w/w*)
Concentration of CMC, C_{CMC}	0.4, 1.0, 1.6	% (*w/w*)
Flow rate of a syringe pump, Q	0.5, 1.0, 1.5	μL/s
Stage speed, v	0.5, 1.0, 1.5	mm/s
Diameter of a nozzle, d	400, 500, 800	μm
Depth of a nozzle, h	5.0, 7.5, 10	mm

To evaluate the corner patterns, a 20 mm line with a single corner (corner angle θ: 30–150°) was printed (ink: 10% (*w/w*) NIPAM monomer, 0.01% (*w/w*) BIS, 1% (*w/w*) New Coccine, and 3% (*w/w*) NaAlg. Conditions: C_{CMC} = 1% (*w/w*), v = 1.0 mm/s, Q = 1.0 μL/s, d = 400 μm, and h = 5 mm) and analyzed. All printing was independently conducted three times. The printed ink was captured when printing using a microscope (VH-5500, KEYENCE, Osaka, Japan) from the z-axis and y-axis.

In all experiments we conducted, we used symbols defined in Table 3. The detailed definitions of these symbols are described in each experiment section.

Table 3. Glossary of symbols.

Define	Symbol
Maximum position of printed ink	z_{top}
Minimum position of printed ink	z_{bottom}
z-axis width of printed ink	w_z
y-axis width of printed ink	w_y
Dragged area of the patterned ink at the corner	A_{error}
Diameter of polymerized ink	$d_{polymer}$
Diameter of printed ink	$d_{initial}$
Diameter of polymerized ink before stimuli	d_0
Diameter of polymerized ink after stimuli	d_n
Patterned ratio of pAAM gel in multi-hydrogel structure	P_A
Patterned ratio of pNIPAMgel in multi-hydrogel structure	P_N
Width of pAAM gel in multi-hydrogel structure	w_A
Width of pNIPAM gel in multi-hydrogel structure	w_N
Total width of multi-hydrogel structure	w_H

2.4. Polymerization of Printed Ink in Supporting Viscous Liquid

Acryloxyethyl thiocarbamoyl rhodamine B (0.002% (*w/w*)) was added to the printing ink to visualize the polymerized hydrogel. After printing a 20 mm straight line of the ink under the standard printing condition (Figure 3b: C_{CMC} = 1% (*w/w*), v = 1.0 mm/s, Q = 1.0 μL/s, d = 400 μm, and h = 5 mm), the printed ink was exposed to UV light (170 mW/cm^2, HLR100T-2, SEN LIGHTS CORPORATION, Osaka, Japan) at an interval time of 20–60 s after printing. After the irradiation, the container of CMC aq was placed into a beaker filled with water to replace CMC with the water to obtain the polymerized ink. The polymerized ink was placed into water that had settled at room temperature, imaged using a fluorescence microscope (IX73P1-22FL/PH, OLYMPUS, Tokyo, Japan), and measured using imaging software (Cellsens, OLYMPUS). Then, we obtained the gelation ratio ($d_{polymer}/d_{initial}$, where $d_{polymer}$ is the diameter of the polymerized ink and $d_{initial}$ is the width of the printed ink).

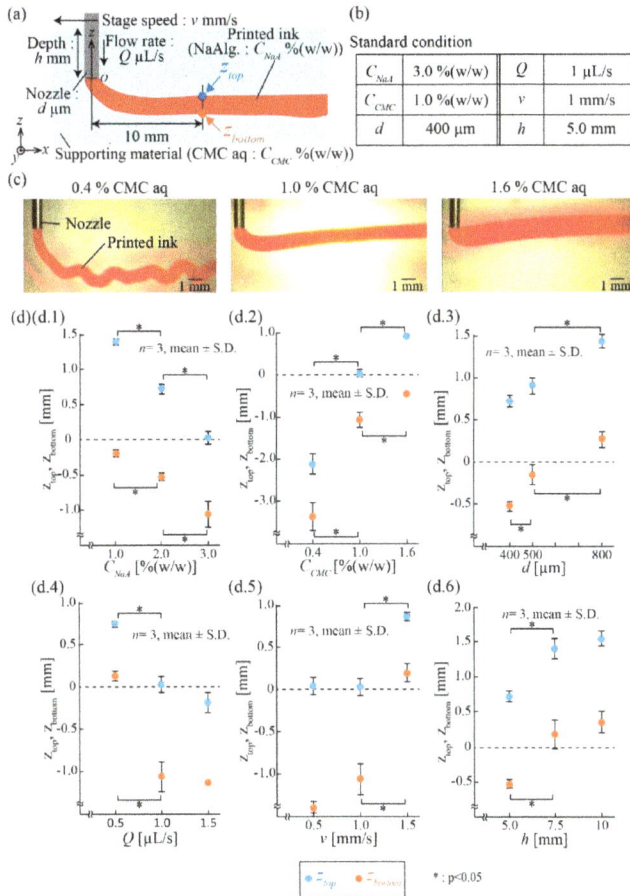

Figure 3. Measurement of the position of the printed ink in the supporting viscous liquid. (**a**) Schematic of z_{top} and z_{bottom} for the printed ink and parameters. (**b**) Table of the standard conditions. (**c**) Images of the printed ink with various concentrations of CMC aq. The z_{top} and z_{bottom} increased when the concentration of CMC aq increased. (**d**) The z_{top} and z_{bottom} for the printed ink under various conditions. The dotted line shows the position of the tip of the nozzle.

2.5. Responsivity of Polymerized Printed Ink

When heating, the polymerized printed ink was placed into heated water (48 °C), which was heated to the specific temperature using a hotplate (ND-1, AS ONE). The temperature of the water was measured using a thermometer (HI98501, Hanna Instruments, Woonsocket, RI, USA). When cooling, the polymerized printed ink was placed into water that had settled at room temperature. The heated or cooled polymerized ink was imaged using a fluorescence microscope and measured using the Cellsens software. Then, we obtained the shrinking ratio (w_n/w_0, where w_n is the diameter of the heated or cooled polymerized ink and w_0 is the initial diameter of the polymerized printed ink).

2.6. Printing of Multi-Hydrogels Structures

We printed multi-hydrogels, including NIPAM-based and AAM-based ink. The NIPAM-based ink was composed of 10% (*w/w*) NIPAM, 0.02% (*w/w*) BIS, 3% (*w/w*) NaAlg, and 1% (*w/w*) fluorescence beads. The AAM-based ink was composed of 10% (*w/w*) acrylamide, 0.02% (*w/w*) BIS, 1% (*w/w*) fluorescence beads, 0.5% (*w/w*) IRGACURE1173, and 3% (*w/w*) NaAlg. We attached a Y-shaped connecter to the nozzle. We printed multi-hydrogels simultaneously, where the flow rate of the NIPAM-based ink was 0.5–0.7 µL/s, and the flow rate of the AAM-based ink was 0.3–0.5 µL/s. UV light was irradiated on the printed ink for 60 s. After the irradiation, the container of CMC aq was placed into a beaker filled with water to get the multi-hydrogel. Then, the multi-hydrogels were placed into heated water (48 °C) for 5 min. We obtained the curvature using a microscope.

2.7. Demonstration of 4D Printing

The NIPAM-based ink was composed of 10% (*w/w*) NIPAM monomer, 0.01% (*w/w*) BIS, 1% (*w/w*) fluorescence beads, 0.5% (*w/w*) IRGACURE1173, and 3% (*w/w*) NaAlg. The AAM-based ink was composed of 10% (*w/w*) acrylamide, 0.02% (*w/w*) BIS, 1% (*w/w*) fluorescence beads, 0.5% (*w/w*) IRGACURE1173, and 3% (*w/w*) NaAlg. We printed a circle, the character "T", and a spring shape using the NIPAM-based ink only. We also printed the character "C" by ejecting NIPAM-based and AAM-based ink. The printed structures were exposed to UV light for 60 s to polymerize. After irradiation, the container of CMC aq was placed into a beaker filled with water to get the polymerized structures. Then, the polymerized structures were placed into water that had settled at room temperature. Then, the polymerized structures were placed into heated water (48 °C), which was heated using a hotplate. The polymerized structures were observed using a microscope.

3. Results

3.1. Position of Printed Ink in Supporting Viscous Liquid

Our printing system was simply composed of a nozzle consisting of a needle, a syringe pump, and *xyz*-motorized precision stages (Figure 2). The printed ink was ejected from the nozzle into a container filled with CMC aq and kept the constant width of the printed pattern. In our system, CMC aq was viscous (8.11–383 mPa·s) and worked as a supporting viscous liquid for the printed ink, enabling us to fabricate 3D hydrogel structures after the polymerization of the patterned ink.

To investigate the printing performance of the ink in CMC aq, we evaluated the printed position of the ink. We printed a straight 20 mm line of ink and measured the top and bottom positions (*z*-axis) of it, z_{top} and z_{bottom}, respectively, at 10 mm from the nozzle (Figure 3a). We tested six parameters: Two liquid parameters (the concentrations of NaAlg in the printed ink C_{NaA} and CMC in the supporting viscous liquid C_{CMC}), two ejection parameters (the diameter *d* of the nozzle and the flow rate *Q* of the ejected ink solution), and two printing parameters (the stage speed *v* and the depth of the nozzle *h*). The listed values for these parameters (Figure 3b) were set as a standard condition for the experiment. These parameters were varied one at a time to examine the effect on the printed ink pattern. For example, Figure 3c presents the images of printed ink patterns in the *z*-axis when C_{CMC} was varied as 0.4%, 1.0%, and 1.6%.

Figure 3d.1–d.6 shows the plots of z_{top} and z_{bottom} when single parameters were varied. For the liquid parameters, z_{top} and z_{bottom} increased when C_{NaA} decreased or C_{CMC} increased (Figure 3d.1,d.2). As these concentrations determine the viscosities of the liquids, the results suggest that the *z* position of the printed ink can be adjusted through the viscosities of the printed ink and supporting fluid. For the ejection parameters, z_{top} and z_{bottom} increased when *d* increased or *Q* decreased (Figure 3d.3,d.4), indicating that the *z* position of the ink depends on the ejection speed, *V*, of the ink at the nozzle tip, described as.

$$V = \frac{Q}{\pi\left(\frac{d}{2}\right)^2} \tag{1}$$

Finally, for the printing parameters, the z_{top} and z_{bottom} increased when *v* or *h* increased (Figure 3d.5,d.6). We consider that pressure loss occurred at the back of the cylindrical nozzle, where the nozzle moved. Because pressure loss depends on *v* and *h*, the position of the printed ink was moved upward.

3.2. Width of Printed Ink in Supporting Viscous Liquid

Next, we investigated the width of the printed ink in terms of the above parameters. Similarly to the experiment in 3.1, we printed a 20 mm straight line of ink and measured the width of the printed ink from the side view (z-axis width), w_z, and top view (y-axis width), w_y, at 10 mm away from the nozzle (Figure 4a,b). Based on the standard conditions in Figure 3b, the parameters were varied one at a time. For example, Figure 4c shows that the widths, w_z and w_y, of the printed ink patterns changed at stage speeds v of 0.5, 1.0, and 1.5 mm/s.

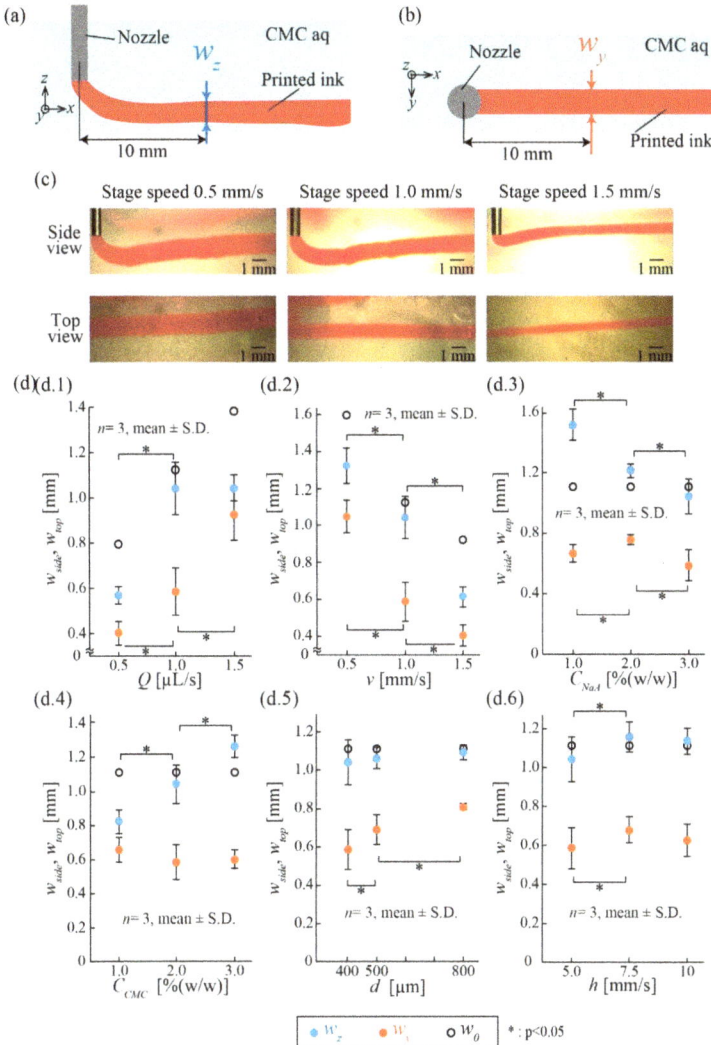

Figure 4. Measurement of the width of the printed ink in the supporting viscous liquid. (**a**) w_z is defined as the z-axis width of the printed ink at 10 mm away from the nozzle. (**b**) w_y is defined as the y-axis width of printed ink at 10 mm away from the nozzle. (**c**) Images of the printed ink with various stage speeds. w_z and w_y decreased as the stage speed increased. (**d**) w_z and w_y under various conditions. w_z and w_y can be mainly controlled by the flow rate and stage speed.

Figure 4d.1–d.6 presents the plots of w_z and w_y when individual parameters were varied. In our experiments, w_z and w_y approximately ranged from 600 μm to 1.3 mm and 400 μm to 1 mm, respectively. In addition to these experimental values, the theoretical width of the printed ink, w_0, expressed as

$$w_0 = \sqrt{\frac{Q}{v\left(\frac{d}{2}\right)^2}} \tag{2}$$

was also plotted as open circles. Note that we hypothesized that the shape of the printed ink was an ideal cylinder.

The widths of the printed ink, w_z and w_y, varied depending on Q and v (Figure 4d.1,d.2). Both widths increased when Q increased or v decreased. This tendency matched with the theoretical description in Equation (1). For the liquid parameters, w_z decreased as C_{NaA} increased, although w_y remained constant (Figure 4d.3). Furthermore, w_z increased as C_{CMC} increased, although w_y also remained constant (Figure 4d.4). We consider that the changes in w_z resulted from the drag force in the z-direction caused by the pressure loss around the nozzle. The two parameters d and h did not influence w_z and w_y (Figure 4d.5,d.6).

The tendencies of the position and width of the printed ink when the parameters varied are summarized in Table 4. The results confirm that, although the z-axis position was affected by the various parameters, the width of the printed ink could be simply controlled by adjusting Q and v. Unless otherwise noted, we adopted the standard conditions (Figure 3b) for the following printing experiments.

Table 4. Trends of z_{top}, z_{bottom}, w_y, and w_z under printing parameter changes.

Parameter	z_{top}, z_{bottom}	w_z	w_y
$C_{NaA}\uparrow$	↓	↑	→
$C_{CMC}\uparrow$	↑	↓	→
$d\uparrow$	↑	→	
$Q\uparrow$	↓		↑
$v\uparrow$	↑		↓
$h\uparrow$	↑		→

3.3. Evaluation of Printed Ink Patterns

The printed inks can be patterned in the supporting liquid using programmed motions of the x, y, and z motors. To examine drawing capability, we printed a line with a single corner of various angles θ ranging from 30 to 150° (Figure 5). Ideally, the patterns of the printed ink should be the same as the tracks of the nozzle. However, the patterns of the printed ink were slightly dragged by the motion of the nozzle and did not exactly match with the tracks of the nozzle (Figure 5a). To evaluate the difference between the printed patterns and the track of the nozzle, we defined the error area A_{error} as the dragged area of the patterned ink at the corner (Figure 5a).

As shown in Figure 5b, the patterned ink ($v = 1.0$ mm/s, $Q = 1.0$ μL/s) was distorted depending on the angle of the corner. We examined the relationship between the error area A_{error} and corner angle θ for three different nozzle speeds and flow rates ($v = 0.5$, 1.0, and 1.5 mm/s: $Q = 0.5$, 1.0, and 1.5 μL/s, respectively) to keep the diameter of the printed ink patterns constant. The error area A_{error} increased as the angle θ increased, reaching the maximum when θ was 120° for all three nozzle speeds (Figure 5c). For all angles θ, the slower the nozzle speed v, the lower the error area A_{error}. According to these results, to print a corner pattern precisely, a small flow rate and slow stage speed should be chosen.

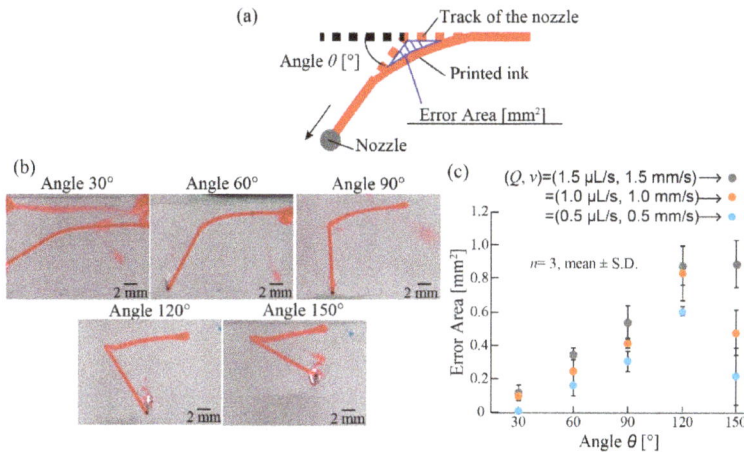

Figure 5. Evaluation of printed ink pattern. (**a**) The error area is defined as the dragged area of the patterned ink at the corner. (**b**) Images of the printed corners with different angles θ. The patterned ink ($v = 1.0$ mm/s, $Q = 1.0$ μL/s) was distorted depending on the angle of the corner. (**c**) Relationship between the error area and θ with different Q and v values. The error area increased as the angle increased, reaching the maximum when the angle was 120° for all three nozzle speeds. For all angles, the slower the nozzle speed, the lower the error area.

3.4. Polymerization of Printed Ink in Supporting Viscous Liquid

In our printing method, the printed ink is polymerized using UV irradiation in the supporting viscous liquid (CMC aq), where the printed ink gradually diffuses. Thus, the interval time between the UV irradiation and printing is important for polymerization. We examined the relationship between the polymerization of the printed ink in CMC aq and the interval time. To verify the polymerization, we defined the gelation ratio as the diameter $d_{polymer}$ of the polymerized pattern after the UV irradiation divided by the diameter $d_{initial}$ of the printed ink pattern before UV irradiation (Figure 6a). Figure 6b shows that the diameter of the polymerized ink became narrower as the interval time increased. Thus, the gelation ratio decreased as the interval time increased (Figure 6c). This is because the gelation area became narrower, owing to the diffusion of the printed ink in CMC aq before UV irradiation. These results indicate that a shorter interval time between printing and UV irradiation can reduce the difference between the polymerized patterns and printed ink patterns.

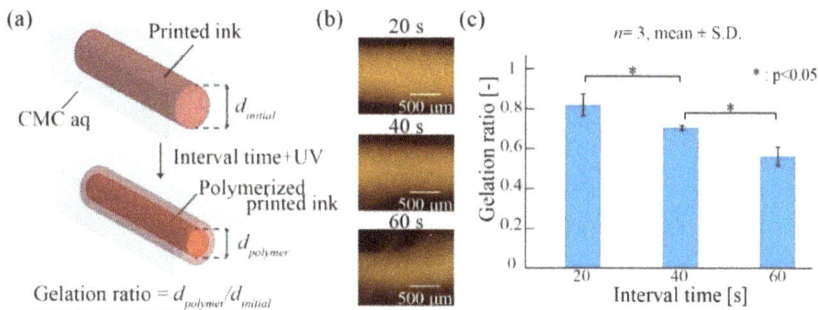

Figure 6. Polymerization of the printed ink in CMC aq. (**a**) Gelation ratio is defined as the diameter of the polymerized ink after UV irradiation divided by the diameter of the printed ink pattern before UV irradiation. (**b**) Fluorescence images of the polymerized printed ink. (**c**) Gelation ratio for different interval times. The gelation ratio decreased as interval time increased.

3.5. Responsivity of External Stimuli

We investigated the responsivity of the polymerized printed hydrogel patterns to thermal stimulation. The shrinking/swelling characteristics of the polymerized hydrogel pattern were evaluated by measuring the shrinking ratio w_n/w_0 in response to changes in the temperature (Figure 7a), where w_0 is the initial width of the polymerized pattern, w_n is the width of the pattern after stimulation, and n is the number of stimuli. We repeatedly heated (48 °C) and cooled (23 °C) the polymerized hydrogel pattern (5 min per each cycle) and evaluated the shrinking ratio (Figure 7b). The shrinking ratio was repeatedly varied from approximately 0.3 to 0.95, corresponding to heating and cooling, respectively. These results suggest that the polymerized printed hydrogel patterns could deform repeatedly in response to heating/cooling cycles.

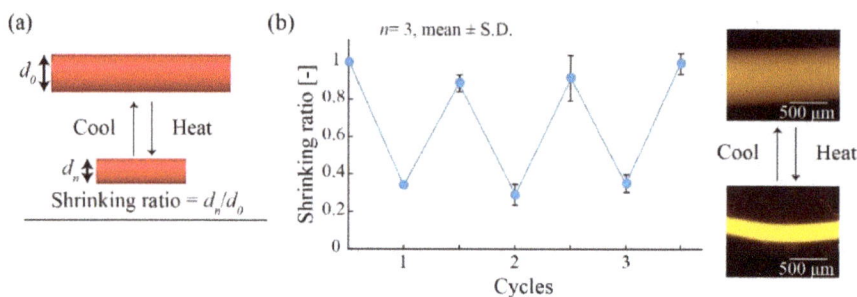

Figure 7. Responsivity of the polymerized hydrogel to thermal stimuli. (**a**) Definition of the shrinking ratio d_n/d_0. Shrinking ratio is obtained by the diameter of the polymerized printed ink after being heated or cooled divided by the initial diameter of the polymerized printed ink before stimuli. (**b**) The repeatability of shrinking/swelling behavior of the polymerized printed ink. The polymerized printed ink shrunk or expanded according to heating or cooling.

3.6. Printing of Multi-Hydrogels Structures

In addition to investigating the simple shrinking/swelling motions of the printed hydrogel patterns, the printing of multiple types of materials was performed, enabling various deformations. We utilized two different pre-gel inks, NIPAM-based and AAM-based, and infused these inks to the nozzle simultaneously via a Y-shaped connecter (Figure 8a). After UV irradiation, the polymerized hydrogel pattern exhibited a double-layer structure. We prepared three different double-layer hydrogel structures which were fabricated by NIPAM/AAM flow rates of 0.5/0.5 µL/s, 0.6/0.4 µL/s, and 0.7/0.3 µL/s. First, we investigated the printing pattern of the double-layer structure by measuring the patterned ratios P_N and P_A ($P_N = w_N/w_H$, $P_A = w_A/w_H$, where w_N is the width of the pNIPAM gel, w_A is the width of the pAAM gel, and w_H is the total width) (Figure 8b). Figure 8c shows that P_N increased and P_A decreased when the flow rate of the NIPAM ink increased and that of the AAM ink decreased. Next, we investigated the responsivity of the double-layer structure to a thermal stimulus. The pNIPAM hydrogel responds to thermal stimuli, whereas the pAAM hydrogel does not. Thus, a folding motion of the polymerized double-layer hydrogel pattern can be achieved in response to a thermal stimulus. We measured the curvatures of these double-layer hydrogel patterns after heating (Figure 8d). The curvature of the double-layer hydrogel structure increased when the flow rate of the NIPAM-based ink increased (Figure 8e,f). A large curvature resulted from increasing the cross-sectional area of the pNIPAM. Figure 8c,e shows that a large curvature can be achieved by setting a large flow rate for the NIPAM ink to increase P_N. These results suggest that deformation control can be achieved by printing multi-hydrogels with controlled layer thicknesses by adjusting the flow rate of the ink.

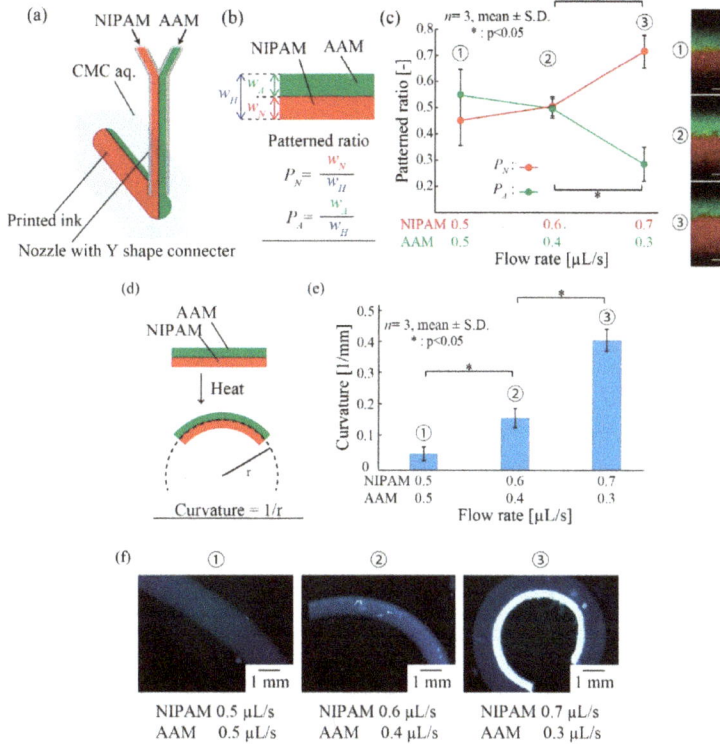

Figure 8. Printing of multi-hydrogel structures. (**a**) A schematic of the setup to print multi-hydrogel inks. The N-isopropylacrylamide-based stimuli-responsive pre-gel solution (NIPAM)-based pre-gel ink and the acrylamide-based non-responsive pre-gel solution (AAM)-based pre-gel ink were ejected into CMC aq via the nozzle attached to a Y shape connector. (**b**) The definition of the patterned ratios P_N and P_A which is obtained by the width of poly-N-isopropylacrylamide (pNIPAM) gel or polyacrylamide (pAAM) gel divided by the total width of the polymerized inks, respectively. (**c**) Results for the patterned ratios P_N and P_A with different flow rates Q. The scale bars are 500 μm. The P_N increased and the P_A decreased when the flow rate of the NIPAM ink increased and that of the AAM ink decreased. (**d**) The definition of the curvature radius of a heated multi-hydrogels structure. (**e**) Results for the curvature radius of heated multi-hydrogels structures with different flow rates. The curvature increased when the flow rate of the NIPAM-based ink increased. (**f**) Images of the deformed multi-hydrogels structures.

3.7. Demonstration of 4D Printing

Finally, we fabricated variations of 4D structures to demonstrate the effectiveness of the proposed method. We set the program of stages to be performed in a circle (Figure 9a.1). Figure 9a.2 presents the images of the fabricated circular structure and its response to thermal stimuli, indicating that not only angulated patterns (see Section 3.3) but also rounded patterns can be fabricated using our method. Furthermore, we printed the character "T" and a 3D spring shape (Figure 9b.1,c.1). Figure 9b.2,c.2 presents the images of the fabricated structures and their response structures, indicating that the cross point (Figure 9b.2) and internal gaps (Figure 9c.2) can be fabricated and maintained following stimulation. By printing cross points, complex 4D structures such as deformable 3D meshes could be fabricated.

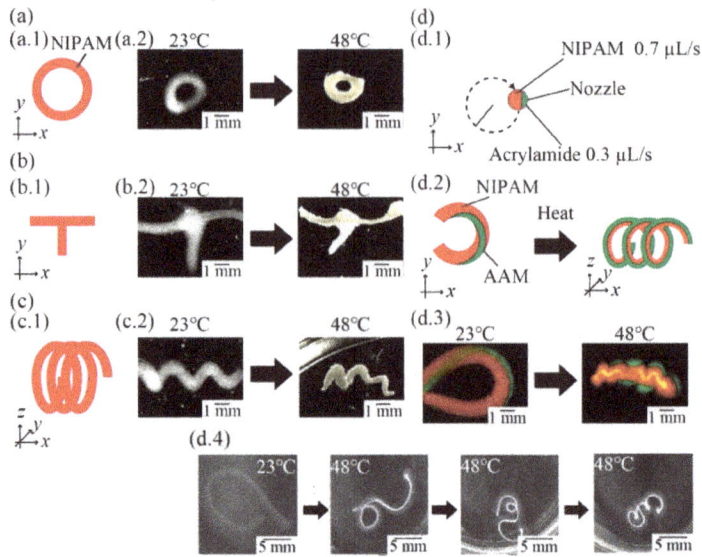

Figure 9. A demonstration of 4D printing. (**a**) Schematic illustration and images of printed structures and heated structures with a rounded pattern. (**b**) Schematic illustration and images of a fabricated T-shaped structure and a heated T-shaped structure with a cross point. (**c**) Schematic illustration and images of a fabricated spring structure and heated spring structure with an internal gap. (**d**) Schematic illustrations of printing a C-shaped structure with multi-hydrogels (**d.1**) and its 3D deformation from the C-shaped structure to the spring-shaped structure. (**d.2**) Fluorescence images of the fabricated C-shaped structure (**d.3**) and the transformed spring structure obtained by heating. (**d.4**) Time-lapse images of the C-shaped structure.

Finally, we printed the character "C" by moving the stages in a "C" shape and ejecting NIPAM-based (0.7 μL/s) and AAM-based (0.3 μL/s) ink at the same time (Figure 9d.1). The printed structure exhibited twisted ink patterns and deformed three-dimensionally into a spiral shape (Figure 9d.2). Figure 9d.3 presents the images of the fabricated structure, a spring shape, and its response. Figure 9d.4 presents the time-lapse images of the deformation. Figure 9d.3,d.4 shows that the fabricated 4D structure can deform three-dimensionally by printing with multi-hydrogels. By printing internal gaps and multi-hydrogels, 4D structures can exhibit various deformations in response to external stimuli, as there is space in which to deform. These results suggest that our proposed method could enable the printing of complex 4D multi-hydrogels structures with cross points and internal gaps.

4. Discussion

We demonstrated a fabrication method for 4D structures composed of multiple types of stimuli-responsive hydrogels while using a viscous liquid as the supporting material during printing with the simple setup: A nozzle, syringe pumps, and motorized stages. Regarding printing performance, the condition shown in Figure 3b is suitable for printing conditions in the current setup. The width of the printed ink is controlled by Q and v rather than the diameter of the nozzle d. In addition, our approach allows us to expand the number of printing materials by using a nozzle with branched channels. Regarding printable structures in our method, it is possible to fabricate 4D structures with internal gaps that have not been fabricated in previous methods [10,11]. By printing internal gaps, it is possible to create functional materials and machines that achieve complex motions or material encapsulation inside the printed structures. Regarding the deformation of 4D structures in response to stimuli with our method, various complex motions of printed structures could be achieved depending on the patterned ratio of multi-layered structures controlled by an adequate flow ratio of NIPAM/AAM inks. In addition, 4D structures with internal gaps can deform in various ways, including the formation of 3D spring structures from a 2D printed C-shaped pattern (Figure 9d) and the anisotropic deformation of multi-layered stimuli-responsive hydrogel springs [24]. Moreover, those 4D structures could deform repeatedly in response to thermal-stimuli because the repeatable responsivity of pNIPAM gel to thermal-stimuli was already reported elsewhere [25,26]. During deformation, the delamination of gel at the cross point (Figure 9b) did not occur (even if the area of the bonding area was smaller than the width of the printed ink) because of proper crosslinking by covalent bond between the layers (c.f. three possible combinations of layers: pNIPAM/pNIPAM gel layers, pNIPAM/pAAM gel layers, and pAAM/pAAM gel layers) [12,27].

Regarding 4D printing techniques, printing ink directly in a viscous liquid is characteristic in our printing method. On the other hand, similar methods in 3D printing techniques have been demonstrated: Printing ink directly in a supporting self-healing gel to fabricate hydrogel 3D structures [28], electronics [29], and vascular networks [30] has been reported. In addition, printing ink directly in granular gel to fabricate silicon 3D structures and hydrogel 3D structures [31] has been reported. Compared to those reported methods, we used a simple viscous supporting material, CMC aq. That is, the printing performance of our method could be enhanced by combining it with reported methods.

There are some challenges in this method. First, the printed ink in the supporting viscous liquid was dragged in the direction of movement of the nozzle because the flow of the supporting viscous liquid (CMC aq) occurred around the nozzle. To improve this, it is necessary to reduce the flow of the supporting material by using a nozzle with a small diameter or by decreasing the stage speed. Second, the printed ink becomes narrow and gradually disappears because of its diffusion in the supporting viscous liquid (Figure 6c). This phenomenon would be a problem for printing large-scale patterns that take longer to print. To solve this problem, it is necessary to polymerize the printed ink immediately after the ink is printed in the supporting viscous liquid. For example, an integrated nozzle that has an optical waveguide for UV irradiation at the tip of the nozzle could be adopted for immediate polymerization.

5. Conclusions

We proposed a new fabrication method for 4D structures composed of stimuli-responsive hydrogels by using a viscous liquid as a supporting material during printing. Using this method, printed 4D structures with internal gaps, which have not been fabricated using previous methods, can be fabricated. We confirmed that the position of the printed ink was influenced by various parameters. The widths of the printed ink, w_z and w_y, ranged from approximately 600 μm to 1.3 mm and 400 μm to 1 mm, respectively. To print 4D structures accurately, a slower stage speed and smaller nozzle diameter should be utilized. Polymerized printed ink that has been irradiated by UV in liquid CMC aq can repeatedly respond to external stimulation. Complex 4D structures exhibiting various deformations in response to external stimuli could be fabricated by printing multi-hydrogels with cross points and internal gaps. We believe that our proposed method would be useful for printing complex 4D structures with multiple functions in environmental monitoring and medical applications.

Author Contributions: T.U. and H.O. conceived and designed the experiments; T.U. performed the experiments and analyzed the data; T.U. and H.O. wrote the paper. All authors discussed the results and contributed to the manuscript.

Funding: This research was funded by Grant-in Aid for Scientific Research (A), Grant No. 18H03868 from Japan Society for the Promotion of Science (JSPS), Japan.

Conflicts of Interest: The authors declare no conflict of interest.

References

1. Momeni, F.; M.Mehdi Hassani.N., S.; Liu, X.; Ni, J. A review of 4D printing. *Mater. Des.* **2017**, *122*, 42–79. [CrossRef]
2. Ding, H.; Zhang, X.; Liu, Y. Review of mechanisms and deformation behaviors in 4D printing. *Int. J. Adv. Manuf. Technol.* **2019**, 1–17. [CrossRef]
3. Kuang, X.; Roach, D.J.; Wu, J.; Hamel, C.M.; Ding, Z.; Wang, T.; Dunn, M.L.; Qi, H.J. Advances in 4D Printing: Materials and Applications. *Adv. Funct. Mater.* **2019**, *29*, 1–23. [CrossRef]
4. Zhang, Z.; Demir, K.G.; Gu, G.X. Developments in 4D-printing: A review on current smart materials, technologies, and applications. *Int. J. Smart Nano Mater.* **2019**, 1–20. [CrossRef]
5. Pei, E. 4D printing: Dawn of an emerging technology cycle. *Assem. Autom.* **2014**, *34*, 310–314. [CrossRef]
6. Zhou, Y.; Huang, W.M.; Kang, S.F.; Wu, X.L.; Lu, H.B.; Fu, J.; Cui, H. From 3D to 4D printing: Approaches and typical applications. *J. Mech. Sci. Technol.* **2015**, *29*, 4281–4288. [CrossRef]
7. Campbell, T.A.; Tibbits, S.; Garrett, B. The Programmable World. *Sci. Am.* **2014**, *311*, 60–65. [CrossRef] [PubMed]
8. Taylor, D.L.; in het Panhuis, M. Self-Healing Hydrogels. *Adv. Mater.* **2016**, *28*, 9060–9093. [CrossRef]
9. Lendlein, A.; Kelch, S. *Shape-Memory Effect*; Cambridge University Press: Cambridge, UK, 1999.
10. Beebe, D.J.; Moore, J.; Bauer, J.M.; Yu, Q.; Liu, R.H.; Devadoss, C.; Jo, B.H. Functional hydrogel structures for autonomous ow control inside micro- uidic channels. *Nature* **2000**, *404*, 588. [CrossRef]
11. Yoon, C.; Xiao, R.; Park, J.; Cha, J.; Nguyen, T.D.; Gracias, D.H. Functional stimuli responsive hydrogel devices by self-folding. *Smart Mater. Struct.* **2014**, *23*, 094008. [CrossRef]
12. Nakajima, S.; Kawano, R.; Onoe, H. Stimuli-responsive hydrogel microfibers with controlled anisotropic shrinkage and cross-sectional geometries. *Soft Matter* **2017**, *13*, 3710–3719. [CrossRef] [PubMed]
13. Ma, C.; Lu, W.; Yang, X.; He, J.; Le, X.; Wang, L.; Zhang, J.; Serpe, M.J.; Huang, Y.; Chen, T. Bioinspired Anisotropic Hydrogel Actuators with On–Off Switchable and Color-Tunable Fluorescence Behaviors. *Adv. Funct. Mater.* **2018**, *28*, 1–7. [CrossRef]
14. Breger, J.C.; Yoon, C.; Xiao, R.; Kwag, H.R.; Wang, M.O.; Fisher, J.P.; Nguyen, T.D.; Gracias, D.H. Self-folding thermo-magnetically responsive soft microgrippers. *ACS Appl. Mater. Interfaces* **2015**, *7*, 3398–3405. [CrossRef] [PubMed]
15. Wang, Z.J.; Zhu, C.N.; Hong, W.; Wu, Z.L.; Zheng, Q. Programmed planar-to-helical shape transformations of composite hydrogels with bioinspired layered fibrous structures. *J. Mater. Chem. B* **2016**, *4*, 7075–7079. [CrossRef]
16. Thérien-Aubin, H.; Wu, Z.L.; Nie, Z.; Kumacheva, E. Multiple shape transformations of composite hydrogel sheets. *J. Am. Chem. Soc.* **2013**, *135*, 4834–4839. [CrossRef] [PubMed]
17. Wu, Z.L.; Moshe, M.; Greener, J.; Therien-Aubin, H.; Nie, Z.; Sharon, E.; Kumacheva, E. Three-dimensional shape transformations of hydrogel sheets induced by small-scale modulation of internal stresses. *Nat. Commun.* **2013**, *4*, 1586–1587. [CrossRef]
18. Zhou, Y.; Hauser, A.W.; Bende, N.P.; Kuzyk, M.G.; Hayward, R.C. Waveguiding Microactuators Based on a Photothermally Responsive Nanocomposite Hydrogel. *Adv. Funct. Mater.* **2016**, *26*, 5447–5452. [CrossRef]
19. Sydney Gladman, A.; Matsumoto, E.A.; Nuzzo, R.G.; Mahadevan, L.; Lewis, J.A. Biomimetic 4D printing. *Nat. Mater.* **2016**, *15*, 413–418. [CrossRef]
20. Naficy, S.; Gately, R.; Gorkin, R.; Xin, H.; Spinks, G.M. 4D Printing of Reversible Shape Morphing Hydrogel Structures. *Macromol. Mater. Eng.* **2017**, *302*, 1–9. [CrossRef]
21. Guo, J.; Zhang, R.; Zhang, L.; Cao, X. 4D printing of robust hydrogels consisted of agarose nanofibers and polyacrylamide. *ACS Macro Lett.* **2018**, *7*, 442–446. [CrossRef]
22. Zhang, M.; Vora, A.; Han, W.; Wojtecki, R.J.; Maune, H.; Le, A.B.A.; Thompson, L.E.; McClelland, G.M.; Ribet, F.; Engler, A.C.; et al. Dual-Responsive Hydrogels for Direct-Write 3D Printing. *Macromolecules* **2015**, *48*, 6482–6488. [CrossRef]
23. Raviv, D.; Zhao, W.; McKnelly, C.; Papadopoulou, A.; Kadambi, A.; Shi, B.; Hirsch, S.; Dikovsky, D.; Zyracki, M.; Olguin, C.; et al. Active printed materials for complex self-evolving deformations. *Sci. Rep.* **2014**, *4*, 1–9. [CrossRef] [PubMed]
24. Yoshida, K.; Nakajima, S.; Kawano, R.; Onoe, H. Spring-shaped stimuli-responsive hydrogel actuator with large deformation. *Sens. Actuators B Chem.* **2018**, *272*, 361–368. [CrossRef]

25. Wang, Y.Q.; Zhang, Y.Y.; Wu, X.G.; He, X.W.; Li, W.Y. Rapid facile in situ synthesis of the Au/Poly(N-isopropylacrylamide) thermosensitive gels as temperature sensors. *Mater. Lett.* **2015**, *143*, 326–329. [CrossRef]

26. Zhang, X.; Pint, C.L.; Lee, M.H.; Schubert, B.E.; Jamshidi, A.; Takei, K.; Ko, H.; Gillies, A.; Bardhan, R.; Urban, J.J.; et al. Optically- and thermally-responsive programmable materials based on carbon nanotube-hydrogel polymer composites. *Nano Lett.* **2011**, *11*, 3239–3244. [CrossRef] [PubMed]

27. Na, J.H.; Evans, A.A.; Bae, J.; Chiappelli, M.C.; Santangelo, C.D.; Lang, R.J.; Hull, T.C.; Hayward, R.C. Programming reversibly self-folding origami with micropatterned photo-crosslinkable polymer trilayers. *Adv. Mater.* **2015**, *27*, 79–85. [CrossRef] [PubMed]

28. Highley, C.B.; Rodell, C.B.; Burdick, J.A. Direct 3D Printing of Shear-Thinning Hydrogels into Self-Healing Hydrogels. *Adv. Mater.* **2015**, *27*, 5075–5079. [CrossRef] [PubMed]

29. Muth, J.T.; Vogt, D.M.; Truby, R.L.; Mengüç, Y.; Kolesky, D.B.; Wood, R.J.; Lewis, J.A. Embedded 3D printing of strain sensors within highly stretchable elastomers. *Adv. Mater.* **2014**, *26*, 6307–6312. [CrossRef] [PubMed]

30. Wu, W.; Deconinck, A.; Lewis, J.A. Omnidirectional printing of 3D microvascular networks. *Adv. Mater.* **2011**, *23*, 178–183. [CrossRef]

31. Bhattacharjee, T.; Zehnder, S.M.; Rowe, K.G.; Jain, S.; Nixon, R.M.; Sawyer, W.G.; Angelini, T.E. Writing in the granular gel medium. *Sci. Adv.* **2015**, *1*, e1500655. [CrossRef]

micromachines

MDPI

Article

Design of Rigidity and Breaking Strain for a Kirigami Structure with Non-Uniform Deformed Regions

Hiroki Taniyama and Eiji Iwase *

Department of Applied Mechanics, Waseda University, 3-4-1 Okubo, Shinjuku-ku, Tokyo 169-8555, Japan;
taniyama@iwaselab.amech.waseda.ac.jp
* Correspondence: iwase@waseda.jp; Tel.: +81-03-5286-2741

Received: 14 May 2019; Accepted: 11 June 2019; Published: 14 June 2019

Abstract: We modeled a kirigami structure by considering the influence of non-uniform deforming cuts in order to theoretically design the mechanical characteristics of the structure. It is known that the end regions of kirigami structures are non-uniformly deformed when stretched, because the deformation is inhibited at the regions close to both the ends connected to the uncut region in the longitudinal direction. The non-uniform deformation affects the overall mechanical characteristics of the structure. Our model was intended to elucidate how cuts at both ends influence these characteristics. We focused on the difference in the deformation degree caused by a cut between the regions close to the ends and the center of the stretched kirigami device. We proposed a model comprising of connected springs in series with different rigidities in the regions close to the ends and the center. The spring model showed good prediction tendency with regard to the curve of the stress–strain diagram obtained using the tensile test with a test piece. Therefore, the results show that it is possible to theoretically design the mechanical characteristics of a kirigami structure, and that such a design can well predict the influence of cuts, which induce non-uniform deformation at both ends.

Keywords: flexible device; stretchable electronic substrate; kirigami structure; mechanical metamaterials

---

1. Introduction

Our objective in this study was to design the mechanical characteristics—such as rigidity and breaking point—of a kirigami structure theoretically by means of a model that considers the influence of cuts, which induce non-uniform deformation at both ends of the stretched kirigami structure.

In recent years, many researchers have conducted investigations using sheets with incised periodic cuts, or so-called kirigami structures. Such structures show interesting mechanical characteristics [1–27]. For example, a kirigami structure can tune the rigidity and breaking strain of the overall device by virtue of changes in its length or the density of cuts in the structure [1–9]. Shyu et al. elucidated the tendency of the stress–strain diagram by stretching the kirigami structure using different parametric conditions and by analyzing the deformations of the structure with finite element modelling [1]. Isobe and Okumura studied the relation between the geometric parameters and rigidity of a kirigami-structured device using the balance of elastic strain energies. They considered a beam model of the kirigami structure and fixed the number of cuts per unit cycle in the longitudinal direction of the structure [2]. Hwang and Bartlett designed a kirigami structure to specifically increase the breaking strain by approximately 1.7 times. The structure was novel in that "minor cuts" were also introduced at both cut ends [3]. Wang et al. developed a kirigami-patterned stretchable conductive film (KSCF) fabricated by laser cutting composites of carbon nanotube conductive networks and an elastic polydimethylsiloxane substrate. Interestingly, the normalized resistance change of the KSCF was 0.10, even after 5000 stretching tests using 0 to 400% strain [4]. Tang and Yin controlled the inclining direction of out-of-plane deformation in a kirigami structure by devising patterns of notches at the front and reverse sides [5]. Lei and Nakatani analyzed the transiting process from in-plane to out-of-plane deformation by modeling a kirigami structure with a beam combination [6]. Rafsanjani and Bertoldi [7] and Tang and Yin [8] designed rigid sheet kirigami structures with square cut units. Moshe et al. proposed an elastic change framework to understand kirigami mechanics in thin sheets with perforations [9]. Dias et al. designed linear actuators that can perform four fundamental forms of linear actuation, that is, roll, pitch, yaw, and lift, by tuning the locations and arrangements of the cuts [10]. Various practical applications have been suggested by changing the mechanical characteristics of kirigami structures, such as increasing the breaking strain of the material and testing steric out-of-plane deformations by stretching the structure [11–26]. Such applications include strain sensors [11–14], stretchable heaters [15], solar cells with solar tracking systems [16], bioprobes [17], crawling robots [18], artificial muscles [19], soft deployable reflectors [20], self-folding hinges [21,22], metamaterial bricks [23], and adhesives with tunable anisotropic adhesive strength [24]. It has been confirmed that a kirigami structure becomes non-uniformly deformed when it is stretched, because the deformation is inhibited at regions close to both ends connected to the uncut region in the longitudinal direction [1–5,11–17]. The non-uniform deforming affects the mechanical characteristics of the overall device. However, the influence of non-uniform deformation on these characteristics has not been analyzed and designed theoretically. In a conventional kirigami structure, the issue of the non-uniform deforming affects has been ignored by increasing the number of patterns per unit cycle of the kirigami structure. This conventional approach can reduce the influence of non-uniform deformation but is not applicable when the range including a kirigami structure is determined or when it is desired to have as low a range as possible. For example, consider a kirigami structure used in an electronic circuit to obtain stretchable wiring parts; it is important to reduce the number of wiring parts to the greatest possible extent and increase the area of the substrate region available for mounting electronic elements. If the mechanical characteristics of the kirigami structure can be theoretically designed while considering non-uniform deformation, it is possible to realize the required mechanical characteristics by making cuts in the minimum range.

In this study, we modeled a kirigami structure to elucidate how cuts at both ends, which induce non-uniform deforming, influence the mechanical characteristics of the overall device. We focused on the difference in the degree of deformation caused by cuts between regions close to the ends and the center of a stretched kirigami device. We proposed a model of connected springs in series based on the hypothesis that there is different rigidity of the regions close to the ends and the center. We derived formulas from the connected springs model, showing the relation between the number of patterns per unit cycle and the mechanical characteristics of the overall device containing the kirigami structure. Comparing the derived model formula and the value measured using the tensile test, we examined whether the proposed model is applicable to kirigami structures. The test piece with a kirigami structure was used for the tensile test. It was made of polyimide (PI) copper (Cu) substrate, which is often used as a film substrate in a flexible device. Section 2 describes the definition of a kirigami structure, and the theory of the spring model proposed in this study and tensile test method. Section 3 describes the mechanical characteristics of a kirigami structure by tensile test. Moreover, we examined whether the proposed spring model was applicable to the kirigami structure by comparing the model values to the spring model values measured by the tensile test. Section 4 describes the conclusions of this study.

2. Theory and Methods

2.1. Definition of a Kirigami Structure

In this paper, we define a kirigami structure as shown in Figure 1. The kirigami structure has periodic cuts in the width direction, and the set of cuts lined in the width direction is repeated in the longitudinal direction. Here, two repetitive cut lines are defined as one pattern cycle. A circular hole is provided at each end of the cut to reduce stress concentration because it is known that when a kirigami structure is stretched, stress is concentrated at both ends of the cut, which could induce cracks [27]. The various dimension parameters that can describe a kirigami structure include cut width w [mm], cut distance d [mm], cut pitch p [mm], diameter of circular hole h [mm], and number of pattern cycles n. The mechanical characteristics of a kirigami structure or a device that includes this structure can change with any modifications to the dimension parameters [1–10]. The kirigami structure shown in Figure 1 is deformed in the three-dimensional and out-of-plane directions after in-plane deformation in two dimensions.

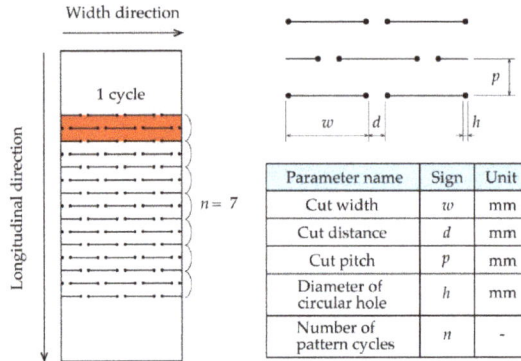

Figure 1. Parameters of a kirigami structure and cut patterns.

2.2. Theory of the Spring Model Proposed in This Study

The stretching of the kirigami structure shown in Figure 1 results in the shape seen in Figure 2a. As shown in Figure 2a, there are differences in deformation between the regions in the center and non-uniformly deformed regions. The deformation in one pattern cycle at the edge is non-uniform compared with the other. Based on this fact, we divided a kirigami structure into three regions, with two regions showing non-uniform deformation at both ends, and one region showing uniform deformation at the center, as shown in Figure 2b. The three divided regions are considered as a model of springs in series, in which two hard springs sandwich $n - 2$ soft springs, as shown in Figure 2c. A typical stress–strain diagram of a kirigami structure shows non-linear curves, as shown in Figure 2d. Thus, we considered elongation stress per strain $E(\varepsilon)$ and breaking strain ε_{cb} to characterize a kirigami structure mechanically. The elongation stress per strain $E(\varepsilon)$ is a physical value that extends Young's modulus beyond the linear region. Therefore, if the structure is in the linear region, $E(\varepsilon)$ is constant and takes the same value as Young's modulus. The definition of the elongation stress per strain is depicted as seen in Equation (1) and Figure 2d.

$$E(\varepsilon) \overset{\text{def}}{=} \frac{\sigma(\varepsilon)}{\varepsilon} \tag{1}$$

Then, we considered that a combined spring, namely the two hard springs and the $n - 2$ soft springs connected in series, are stretched by stress σ, as shown in Figure 2c. We assumed that the strains of the two hard springs and the $n - 2$ soft springs are ε_0 and ε_1, respectively, and the combined strain of the whole spring of length L is ε_c. Considering the displacement by stress σ_c, we obtained Equation (2).

$$\varepsilon_c L = \frac{2}{n}\varepsilon_0 L + \frac{n - 2}{n}\varepsilon_1 L \tag{2}$$

In Equation (2), we assumed that each spring has the same length (L/n), corresponding to the cut pitch of the kirigami structure. Then, we posited that the two hard springs and the $n - 2$ soft springs have an elongation stress per strain $E_0(\varepsilon_0)$ and $E_1(\varepsilon_1)$, respectively, as shown in Figure 2c. Considering the balance of stress when the springs connected in series are stretched by stress σ, the combined elongation stress per strain of the whole spring $E_c(\varepsilon_c)$ is represented as

$$\varepsilon_c E_c(\varepsilon_c) = \varepsilon_0 E_0(\varepsilon_0) = \varepsilon_1 E_1(\varepsilon_1) \tag{3}$$

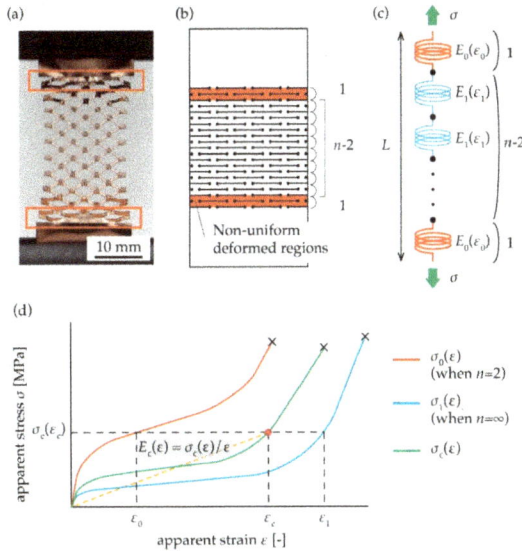

Figure 2. (**a**) Photograph of a stretched kirigami structure. (**b**) Definition of non-uniform deformed regions at both ends. (**c**) A spring model of the kirigami structure, in which two hard springs with an elongation stress per strain $E_0(\varepsilon_0)$ sandwich $n - 2$ soft springs with an elongation stress per strain $E_1(\varepsilon_1)$. (**d**) The combined stress–strain curve of the whole spring $\sigma_c(\varepsilon)$ can be obtained from stress–strain curves $\sigma_0(\varepsilon)$ and $\sigma_1(\varepsilon)$ of each spring.

In Equation (3), we assumed that all the springs are of the same cross-sectional area, corresponding to the cross-sectional area of the substrate. From Equations (2) and (3), the elongation stresses per strain of each spring $E_0(\varepsilon_0)$ and $E_1(\varepsilon_1)$ are used to express the elongation stress per strain of the whole spring $E_c(\varepsilon_c)$, as shown in Equation (4) and Figure 2d.

$$E_c(\varepsilon_c) = \frac{n E_0(\varepsilon_0) E_1(\varepsilon_1)}{2 E_1(\varepsilon_1) + (n-2) E_0(\varepsilon_0)} \tag{4}$$

We used the model of Equation (4) to analyze the kirigami in Figure 1. In our analysis, if we obtained $E_0(\varepsilon_0)$ and $E_1(\varepsilon_1)$ experimentally, we could obtain the elongation stress per strain $E_c(\varepsilon_c)$ for any number of pattern cycles n. The values $E_0(\varepsilon_0)$ and $E_1(\varepsilon_1)$ were obtained by fitting to the model of Equation (4) using the measured value $E_c(\varepsilon_c)$ on various number of pattern cycles n. In addition, we considered the connected springs in series model to elucidate the relationship between n and breaking strain. The breaking strains of the springs with the respective elongation stress per strain $E_0(\varepsilon_0)$ and $E_1(\varepsilon_1)$ are defined as ε_{0b} and ε_{1b} respectively. If we assume the breaking strains ε_{0b} and ε_{1b} are almost the same, we can obtain the combined breaking strain ε_{cb} of the whole spring is as follows from Equation (2).

$$\varepsilon_{cb} = \frac{2}{n} \varepsilon_{0b} + \frac{n-2}{n} \varepsilon_{1b} \tag{5}$$

In our analysis, the values ε_{0b} and ε_{1b} were obtained by fitting to the model of Equation (5) using the measured values ε_{cb} on various number of pattern cycles n.

2.3. Tensile Testing Method

Comparing the derived formula and results from the tensile test, we examined the applicability of the proposed model to the kirigami structure. Test pieces for the tensile test were fabricated using PI-Cu substrate (Toray Advanced Materials Korea, Metaroiyal®), and a film of Cu was formed by sputtering and electrolytic Cu plating. The thicknesses of the PI and Cu layers on the substrate were 25 μm and 8 μm, respectively. The test piece for the tensile test was shaped as a bar and the kirigami structure was included in the central region. The test piece included non-cut areas, which measured 10 mm in the longitudinal direction, at both ends. The kirigami structure for the test piece was cut out using an ultraviolet laser beam machine (Osada Photonics International, OLMUV-355-5A-K) with a wavelength of 355 nm. The tensile test was performed by stretching the test piece using the tensile testing machine with an integrated force gauge (IMADA, ZTA-5N) and electric measurement stand (IMADA, MX2-500N). The test piece was fixated by jigs (IMADA, FC-41U, FC-41U-F) equipped with urethane rubber on one side, so as not to break the chuck region on stretching. Figure 3 shows the experimental setup for the tensile test. The test piece was fixed by the jig so that the tested length was 6 mm from both ends in the longitudinal direction. The static tensile test was selected in order to evaluate the mechanical characteristics of the kirigami structure, and the lifting speed of the tensile test was 10 mm/min, which was sufficiently slow compared to the length of the test piece. When stretching the test piece with the tensile testing machine, load and elongation were monitored by the force gauge, and the load–elongation diagram was prepared for the kirigami structure. The test piece started stretching from its bended state and, once the load reached 0.01 N, the zero position of load–elongation diagram was set (namely, the load and elongation were zero at this point). The stress–strain diagram was produced using the obtained load–elongation diagram. The obtained load was divided by the cross-sectional area of the substrate, and the strain was divided by the gauge length of the test piece. Here, we considered cross-sectional area of the substrate without cuts and defined length between the ends of the cuts in the kirigami structure as the gauge length of the test piece. Figure 3b shows definition of the cross-sectional area and gauge length of the test piece. The dimension parameters of the test piece were fixed as follows: cut width w, cut distance d, cut pitch p, diameter of each round hole h, and thickness of the substrate were 5 mm, 1 mm, 1 mm, 0.3 mm, and 33 μm, respectively. Six types of pattern cycles (n) were used (3, 5, 10, 15, 20, and 25). A tensile test was conducted five times for each of the six patterns.

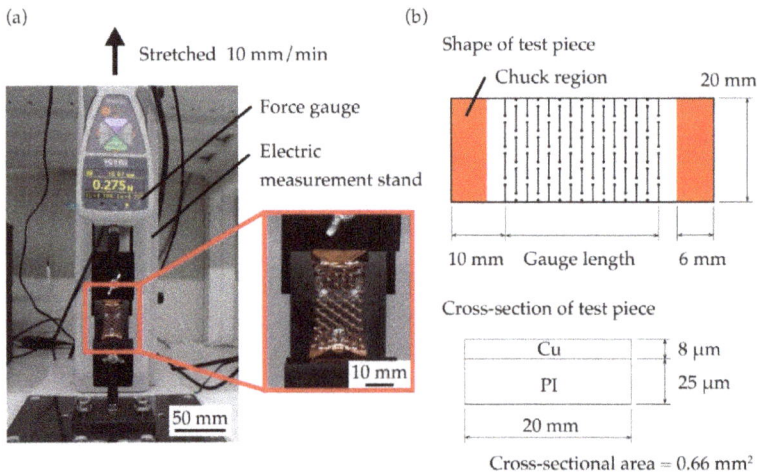

Figure 3. (**a**) Photograph of the tensile test setup, and (**b**) definition of the test piece shape in the tensile test and parameter dimensions.

3. Results and Discussion

We examined whether the proposed spring model was applicable to the kirigami structure by comparing the measured values when n was changed using Equations (4) and (5). Figure 4 shows the stress–strain diagram using the tensile test. When n was 25, the stroke reached the upper limit of the electric measurement device, and the test piece did not break. The elongation stress per strain and breaking strain were calculated using the stress–strain diagram in Figure 4. Figure 5 shows the relationship between the mechanical characteristics of the kirigami structure and n. As shown in Figure 5a, elongation stress per strain decreased as n increased. As shown in Figure 5b, breaking strain also increased as n increased. In the case of $n = 10$ where the influence of the non-uniformed cuts at both ends is large, $E_c(\varepsilon_c)$ increased by about 20% and ε_{cb} decreased by about 15%, compared to the case of $n = \infty$. The case of $n = \infty$ means the ideal case occurs without the influence of the non-uniformed cuts at both ends. These graphs indicated that the increase in n reduced the influence of non-uniform deformation in both end regions, and the uniform deformation in the center region became predominant toward determining the elongation stress per strain and breaking strain of the overall device. The values of $E_0(\varepsilon_0)$, $E_1(\varepsilon_1)$, ε_{0b}, and ε_{1b} were then calculated from the measured values $E_c(\varepsilon_c)$, ε_{0b}, and ε_{1b} on various number of pattern cycles n as shown in Figure 5. The fitting curves using the least squares method in Figure 5 were produced using Equations (4) and (5). In Figure 5, we show that the values of the elongation stress per strain are $E_0(\varepsilon_0) = 32.8$ MPa and $E_1(\varepsilon_1) = 0.878$ MPa at $\sigma_c = 1.0$ MPa, $E_0(\varepsilon_0) = 7.78$ MPa and $E_1(\varepsilon_1) = 1.49$ MPa at $\sigma_c = 2.0$ MPa, $E_0(\varepsilon_0) = 5.83$ MPa and $E_1(\varepsilon_1) = 1.78$ MPa at $\sigma_c = 3.0$ MPa. In addition, we found that the values of the breaking strain were $\varepsilon_{0b} = 0.451$ strain, and $\varepsilon_{1b} = 1.41$ strain. It was found that the proposed spring model was applicable to the kirigami structure, because all the measured values fit the theoretical curves of Equations (4) and (5). We confirmed whether the predicted curve of the stress–strain diagram from the spring model shows the mechanical characteristics of the overall kirigami-structured device. When $n = 10$, the non-uniformly deformed regions at both ends predominantly influence the mechanical characteristics of the overall device. The average values of the elongation stress per strain and breaking strain of the overall device at $n = 10$ in Figure 4 were plotted in the stress–strain diagram in Figure 6. The measured values of $E_0(\varepsilon_0)$, $E_1(\varepsilon_1)$, ε_{0b}, and ε_{1b} for 10 pattern cycles were substituted into Equations (4) and (5). Based on these equations, the model values of the elongation stress per strain and breaking strain for $n = 10$ were plotted in the stress–strain diagram in Figure 6. The values derived from the equations and the measured values were compared and found to be in good agreement. Accordingly, theoretical design of the mechanical characteristics of the kirigami structure was made possible by applying the spring model to the kirigami structure; the model can well predict the influence of cuts, which induce non-uniform deformation at both ends.

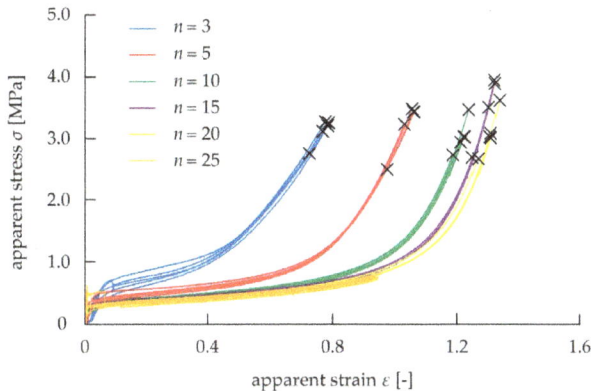

Figure 4. Stress–strain diagram when stretching the kirigami structure for different values of n.

(a)

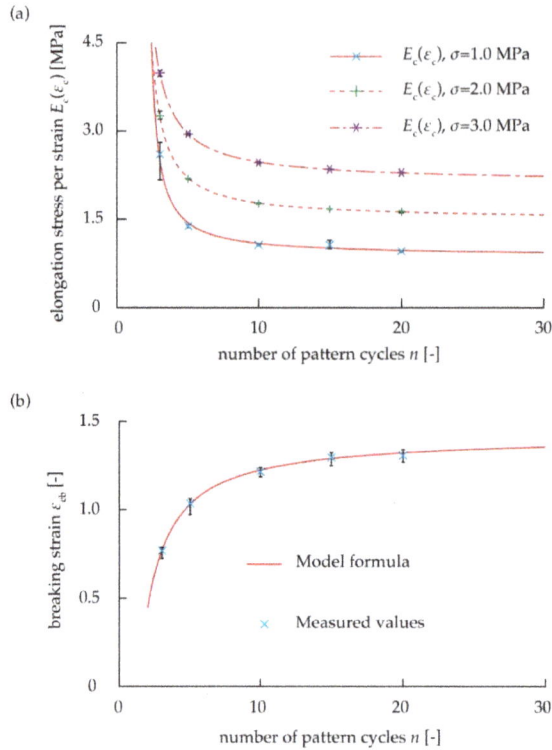

(b)

Figure 5. Comparison between fitting curve by theoretical and measured values: (**a**) Relation between n and elongation stress per strain, (**b**) Relationship between n and breaking strain.

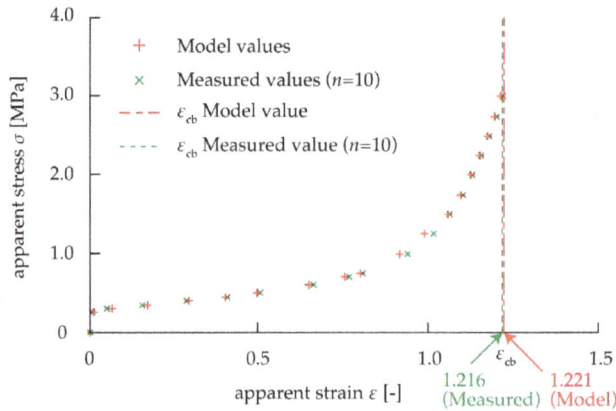

Figure 6. Comparison of values calculated using the spring model and values in the stress–strain diagram from the tensile test for $n = 10$ using the test piece.

4. Conclusions

In order to design the mechanical characteristics of a kirigami structure theoretically, we modeled the stretching of such a structure when considering the influence of cuts inducing non-uniform deformation at both ends. The stress–strain diagram for the stretched kirigami structure showed non-linear curves. Thus, we considered elongation stress per strain and breaking strain to characterize the kirigami structure mechanically. For the modeling, we focused on the difference in the deformation degree of the cut and divided the kirigami structure into three regions: two non-uniform deformed regions for one cycle at both ends, and one uniform deformed region at the center. We proposed a model of connected springs in series, which had different elongation stress per strain depending on the regions, considering the cuts in one cycle as one spring. We derived formulas from the spring model, which showed the dependence of elongation stress per strain and breaking strain of the kirigami structure on the number of pattern cycles n. Comparing the derived formulas and the measured values, we validated the proposed model. The values provided by the formulas of the spring model were in good agreement with the values measured from the tensile test for both elongation stress per strain and breaking strain, illustrating that the proposed spring model was applicable to the kirigami structure. Our proposed spring model well predicted the tendency of the stress–strain diagram curve of the kirigami structure test piece with $n = 10$, and compared to case of $n = \infty$, it showed the large influence of the non-uniform deformation at both ends, $E_c(\varepsilon_c)$ increased by about 20% and ε_{cb} decreased by about 15%. Based on the above, it is possible to theoretically design the mechanical characteristics of a kirigami structure for certain values of w, d and p, while considering the influence of non-uniformly deformed cuts at both ends. This can be accomplished by calculating $E_0(\varepsilon_0)$, $E_1(\varepsilon_1)$, ε_{0b}, and ε_{1b} in Equations (4) and (5) by fitting from measured values.

Author Contributions: H.T. and E.I. conceived and designed the experiments; H.T. performed the experiments, analyzed the data, and wrote the paper; E.I. supervised the research.

Funding: This work was partially supported by JSPS KAKENHI Grant Number 18H03868.

Acknowledgments: The authors would like to acknowledge Hiroki Yasuga of Waseda University for his support.

Conflicts of Interest: The authors declare no conflict of interest.

References

1. Shyu, T.C.; Damasceno, P.F.; Dodd, P.M.; Lamoureux, A.; Xu, L.; Shlian, M.; Shtein, M.; Glotzer, S.C.; Kotov, N.A. A kirigami approach to engineering elasticity in nanocomposites through patterned defects. *Nat. Mater.* **2015**, *14*, 785–789. [CrossRef] [PubMed]
2. Isobe, M.; Okumura, K. Initial rigid response and softening transition of highly stretchable kirigami sheet materials. *Sci. Rep.* **2016**, *6*, 1–6. [CrossRef] [PubMed]
3. Hwang, D.G.; Bartlett, M.D. Tunable mechanical metamaterials through hybrid kirigami structures. *Sci. Rep.* **2018**, *8*, 1–8. [CrossRef] [PubMed]
4. Wang, Z.; Zhang, L.; Duan, S.; Jiang, H.; Shen, J.; Li, C. Kirigami-patterned highly stretchable conductors from flexible carbon nanotube-embedded polymer films. *J. Mater. Chem. C* **2017**, *5*, 8714–8722. [CrossRef]
5. Tang, Y.; Lin, G.; Yang, S.; Yi, Y.K.; Kamien, R.D.; Yin, J. Programmable kiri-kirigami metamaterials. *Adv. Mater.* **2017**, *29*, 1–9. [CrossRef]
6. Lei, X.-W.; Nakatani, A.; Doi, Y.; Matsunaga, S. Bifurcation analysis of periodic kirigami structure with out-plane deformation. *J. Soc. Mater. Sci. Jpn.* **2018**, *67*, 202–207. [CrossRef]
7. Rafsanjani, A.; Bertoldi, K. Buckling-induced kirigami. *Phys. Rev. Lett.* **2017**, *118*, 1–5. [CrossRef]
8. Tang, Y.; Yin, J. Design of cut unit geometry in hierarchical kirigami-based auxetic metamaterials for high stretchability and compressibility. *Extrem. Mech. Lett.* **2017**, *12*, 77–85. [CrossRef]
9. Moshe, M.; Esposito, E.; Shankar, S.; Bircan, B.; Cohen, I.; Nelson, D.R.; Bowick, M.J. Kirigami mechanics as stress relief by elastic charges. *Phys. Rev. Lett.* **2019**, *122*, 048001. [CrossRef]
10. Dias, M.A.; McCarron, M.P.; Rayneau-Kirkhope, D.; Hanakata, P.Z.; Campbell, D.K.; Park, H.S.; Holmes, D.P. Kirigami actuators. *Soft Matter* **2017**, *13*, 9087–9092. [CrossRef]
11. Zheng, W.; Huang, W.; Gao, F.; Yang, H.; Dai, M.; Liu, G.; Yang, B.; Zhang, J.; Fu, Y.Q.; Chen, X.; et al. Kirigami-inspired highly stretchable nanoscale devices using multidimensional deformation of monolayer MoS₂. *Chem. Mater.* **2018**, *30*, 6063–6070. [CrossRef]
12. Sun, R.; Zhang, B.; Yang, L.; Zhang, W.; Farrow, I.; Scarpa, F.; Rossiter, J. Kirigami stretchable strain sensors with enhanced piezoelectricity induced by topological electrodes. *Appl. Phys. Lett.* **2018**, *112*, 1–6. [CrossRef]

13. Gao, B.; Elbaz, A.; He, Z.; Xie, Z.; Xu, H.; Liu, S.; Su, E.; Liu, H.; Gu, Z. Bioinspired kirigami fish-based highly stretched wearable biosensor for human biochemical–physiological hybrid monitoring. *Adv. Mater. Technol.* **2018**, *3*, 1–8. [CrossRef]

14. Baldwin, A.; Meng, E. A kirigami-based Parylene C stretch sensor. In Proceedings of the 2017 IEEE 30th International Conference on Micro Electro Mechanical Systems (MEMS), Las Vegas, NV, USA, 22–26 January 2017; pp. 227–230. [CrossRef]

15. Jang, N.S.; Kim, K.H.; Ha, S.H.; Jung, S.H.; Lee, H.M.; Kim, J.M. Simple approach to high-performance stretchable heaters based on kirigami patterning of conductive paper for wearable thermotherapy applications. *ACS Appl. Mater. Interfaces* **2017**, *9*, 19612–19621. [CrossRef] [PubMed]

16. Lamoureux, A.; Lee, K.; Shlian, M.; Forrest, S.R.; Shtein, M. Dynamic kirigami structures for integrated solar tracking. *Nat. Commun.* **2015**, *6*, 1–6. [CrossRef]

17. Morikawa, Y.; Yamagiwa, S.; Sawahata, H.; Numano, R.; Koida, K.; Ishida, M.; Kawano, T. Ultrastretchable kirigami bioprobes. *Adv. Healthc. Mater.* **2018**, *7*, 1–10. [CrossRef]

18. Lee, D.; Saito, K.; Umedachi, T.; Ta, T.D.; Kawahara, Y. Origami robots with flexible printed circuit sheets. In Proceedings of the 2018 ACM International Joint Conference and 2018 International Symposium on Pervasive and Ubiquitous Computing and Wearable Computers, Singapore, 8–12 October 2018; pp. 392–395. [CrossRef]

19. Sareh, S.; Rossiter, J. Kirigami artificial muscles with complex biologically inspired morphologies. *Smart Mater. Struct.* **2013**, *22*. [CrossRef]

20. Wang, W.; Li, C.; Rodrigue, H.; Yuan, F.; Han, M.W.; Cho, M.; Ahn, S.H. Kirigami/origami-based soft deployable reflector for optical beam steering. *Adv. Funct. Mater.* **2017**, *27*, 1–9. [CrossRef]

21. Zhang, Q.; Wommer, J.; O'Rourke, C.; Teitelman, J.; Tang, Y.; Robison, J.; Lin, G.; Yin, J. Origami and kirigami inspired self-folding for programming three-dimensional shape shifting of polymer sheets with light. *Extrem. Mech. Lett.* **2017**, *11*, 111–120. [CrossRef]

22. Kwok, T.-H.; Wang, C.C.L.; Deng, D.; Zhang, Y.; Chen, Y. Four-dimensional printing for freeform surfaces: Design optimization of origami and kirigami structures. *J. Mech. Des.* **2015**, *137*, 111413. [CrossRef]

23. Memoli, G.; Caleap, M.; Asakawa, M.; Sahoo, D.R.; Drinkwater, B.W.; Subramanian, S. Metamaterial bricks and quantization of meta-surfaces. *Nat. Commun.* **2017**, *8*, 1–8. [CrossRef] [PubMed]

24. Hwang, D.G.; Trent, K.; Bartlett, M.D. Kirigami-inspired structures for smart adhesion. *ACS Appl. Mater. Interfaces* **2018**, *10*, 6747–6754. [CrossRef] [PubMed]

25. Xu, L.; Shyu, T.C.; Kotov, N.A. Origami and kirigami nanocomposites. *ACS Nano* **2017**, *11*, 7587–7599. [CrossRef] [PubMed]

26. Ning, X.; Wang, X.; Zhang, Y.; Yu, X.; Choi, D.; Zheng, N.; Kim, D.S.; Huang, Y.; Zhang, Y.; Rogers, J.A. Assembly of advanced materials into 3D functional structures by methods inspired by origami and kirigami: A review. *Adv. Mater. Interfaces* **2018**, *5*, 1–13. [CrossRef]

27. Chen, S.H.; Chan, K.C.; Yue, T.M.; Wu, F.F. Highly stretchable kirigami metallic glass structures with ultra-small strain energy loss. *Scr. Mater.* **2018**, *142*, 83–87. [CrossRef]

micromachines

MDPI

Article

Resistance Change Mechanism of Electronic Component Mounting through Contact Pressure Using Elastic Adhesive

Takashi Sato [1], Tomoya Koshi [2] and Eiji Iwase [1],*

[1] Department of Applied Mechanics, Waseda University, 3-4-1 Okubo, Shinjuku-ku, Tokyo 169-8555, Japan; sato@iwaselab.amech.waseda.ac.jp

[2] Sensing System Research Center, National Institute of Advanced Industrial Science and Technology (AIST), 1-1-1 Higashi, Tsukuba 305-8565, Japan; t.koshi@aist.go.jp

* Correspondence: iwase@waseda.jp; Tel.: +81-03-5286-2741

Received: 19 May 2019; Accepted: 10 June 2019; Published: 14 June 2019

Abstract: For mounting electronic components through contact pressure using elastic adhesives, a high contact resistance is an inevitable issue in achieving solderless wiring in a low-temperature and low-cost process. To decrease the contact resistance, we investigated the resistance change mechanism by measuring the contact resistance with various contact pressures and copper layer thicknesses. The contact resistivity decreased to 4.2×10^{-8} $\Omega \cdot m^2$ as the contact pressure increased to 800 kPa and the copper layer thickness decreased to 5 µm. In addition, we measured the change in the total resistance with various copper layer thicknesses, including the contact and wiring resistance, and obtained the minimum combined resistance of 123 mΩ with a copper-layer thickness of 30 µm using our mounting method. In this measurement, a low contact resistance was obtained with a 5-µm-thick copper layer and a contact pressure of 200 kPa or more; however, there is a trade-off with respect to the copper layer thickness in obtaining the minimum combined resistance because of the increasing wiring resistance. Subsequently, based on these measurements, we developed a sandwich structure to decrease the contact resistance, and a contact resistivity of 8.0×10^{-8} $\Omega \cdot m^2$ was obtained with the proposed structure.

Keywords: surface mounting; flexible electronic device; contact resistance; contact pressure

1. Introduction

Recently, flexible electronic devices, such as flexible displays [1–5], batteries [6–10], and sensor arrays [11–15], have been developed by many research groups [16–22]. In particular, healthcare monitoring systems using flexible electronic devices, which can adhere to human skin, have attracted considerable interest. In previous studies, to mount electronic components on a flexible circuit, solders [23,24] or conductive adhesives [25,26] were used, and the contact resistivity is on the order of 10^{-11} $\Omega \cdot m^2$ in solders and 10^{-9} $\Omega \cdot m^2$ in conductive adhesives. The conductivity is the contact resistance per unit contact area. Although a simple fabrication process is required for high-mix, low-volume manufacturing to achieve individual optimization of healthcare monitoring systems, these fabrication processes are quite complicated owing to both the dispensing and patterning processes. Moreover, polyurethane and rubbers are mainly used for substrates of stretchable circuits, but their low heat resistance causes problems in soldering and misalignment of electronic components owing to thermal expansion. To solve these issues, solderless mounting methods via contact pressure using an elastic adhesive [27,28] were proposed as a new simple manufacturing processes. These methods do not require complex mounting processes and heating. Their applicability is, however, hindered by their high contact resistance because the electrical connections are based on physical contact. The mechanism for controlling resistance change to achieve low contact resistance is still not clear.

Therefore, in this study, we first considered the mechanism of contact resistance change to clarify the causes of this change. Then, the change in the contact resistance was experimentally measured to investigate the relationship between the expected causes and the contact resistance. Second, we developed a new mounting structure to obtain a high compression force to decrease the contact resistance and evaluated the contact resistance.

2. Materials and Methods

First, the basic mechanism of contact resistance was considered. Contact resistance is the sum of the constriction resistance and film resistance. The constriction resistance is influenced by the concentration of current that flows into the real contact area between the conductor surfaces. The film resistance is influenced by chemical films, such as oxide films, oil, and dirt films, on the conductor surface. In this research, we focused on constriction resistance because the constriction resistance must be dominant in this solderless mounting of electronic components; constriction resistance would decrease as the real contact area increased. Figure 1a shows the schematic image of the contact resistance change in the solderless mounting method. The electrical connection is formed via contact of surface-mounted electronic components with contact pads of the metal layer, where the elastic adhesives provide restoring forces to press them. For example, a surface-mounted electronic component is held by elastic adhesives sandwiching from above and below in the sandwich structure proposed in this research, shown in Figure 1c. The important feature here is the surface shape of the electrode of the electronic components, as shown in Figure 1b. Most of the electrodes in surface-mounted electronic components are generally not perfectly flat but slightly curved and rough, and so are the chip resistors. These electrodes are gradual convex shapes 400-μm wide and 20-μm high, and the surface roughness of the electrodes is less than a few micrometers. Figure 1 shows the decrease in the contact resistance caused by the change in the contact area when an electronic component is pressed against the contact pads. On applying the contact pressure, the contact pad deforms and the contact area with the electrodes of the component is enlarged. Hence, Figure 1c suggests that the thinner metal electrodes exhibit large deformation, and thus they make contact with the electrodes to lower the contact resistance. Nevertheless, the thinner metal layers show larger resistance and are susceptible to damage owing to excessive deformation. Therefore, it is important to investigate contact pressures to obtain a sufficiently small contact resistance with different metal layer thicknesses and determine the thickness when the sum of the contact resistance and wiring resistance is the lowest.

To evaluate the effect of the contact pressure and thickness of the metal layer on the contact resistance, we experimentally measured the change in the contact resistance considering the contact pressure and thickness of the metal layer. Figure 2a shows schematic images with the dimensions of the experimental sample. The acrylic foam double sided adhesive sheet (Y-4905J, 3M, Maplewood, MN, USA) with a thickness of 0.5 mm was cut to a length of approximately 30 mm and width of 20 mm and affixed to a glass plate. A metal layer was patterned by laser cutting to a length of 10 mm and width of 5 mm using rolled copper films (Nilaco Co., Tokyo, Japan) with thicknesses of 5, 10, 30, and 50 μm. The metal layers were transferred onto the elastic adhesive with a gap of 1 mm, and a surface-mounted chip resistor (MCR10EZPJ000, ROHM Co., Kyoto, Japan) was placed on it. Figure 2b shows optical images of a sample used in the experiment. Figure 3a,b shows optical images of the apparatus used to measure the change in the contact resistance. The sample was fixed onto a movable stage with polyimide tape to be pushed by a compression testing machine (ZTS-20N, IMADA Co., Aichi, Japan); this machine has a pole with a small and flat end to press the chip resistor. The movable stage went up to apply pressure on the chip resistor, and the value of the pressure was monitored by the compression testing machine. The contact resistance was measured using a source meter (2614B, Keithley Instruments, Cleveland, OH, USA) with four-terminal sensing. The applied voltage was 10 mV, and the current compliance was 100 mA. Figure 3c shows the backside images of the experimental sample without and with pushing by a compression testing machine. With pressing pressure, the chip resistor sunk into the contact pad regions of the metal layer, and the contact pads largely deformed along the surface of the chip resistor. To measure and compare the total resistance, contact resistance ($R_{contact}$), and wiring resistance of the metal layer (R_{wiring}), we fabricated a sample device composed of a surface-mounted electronic component with a metal layer of contact pads and wave-shaped wirings and an elastic adhesive (Figure 4). We employed a chip resistor (MCR10EZPJ000, ROHM Co., Kyoto, Japan) with an internal resistance ($R_{internal}$) of 11 mΩ, a 0.5-mm-thick acrylic foam adhesive sheet, and a metal layer of rolled copper films with thicknesses of 5, 10, 30, and 50 μm for the device fabrication. The copper films were patterned to form contact pads and wave-shaped wirings with a width of 1.0 mm and radius of 0.9 mm. The contact pads were fabricated with a length of 2.8 mm and width of 2.8 mm at both ends of the wiring. The patterned copper film was placed on the adhesive sheet with a gap of 1 mm. R_{wiring} of the wave-shaped electrodes was measured by the source meter with four-terminal sensing. Total resistance (R_{total}) is the sum of $R_{contact}$, R_{wiring}, and $R_{internal}$.

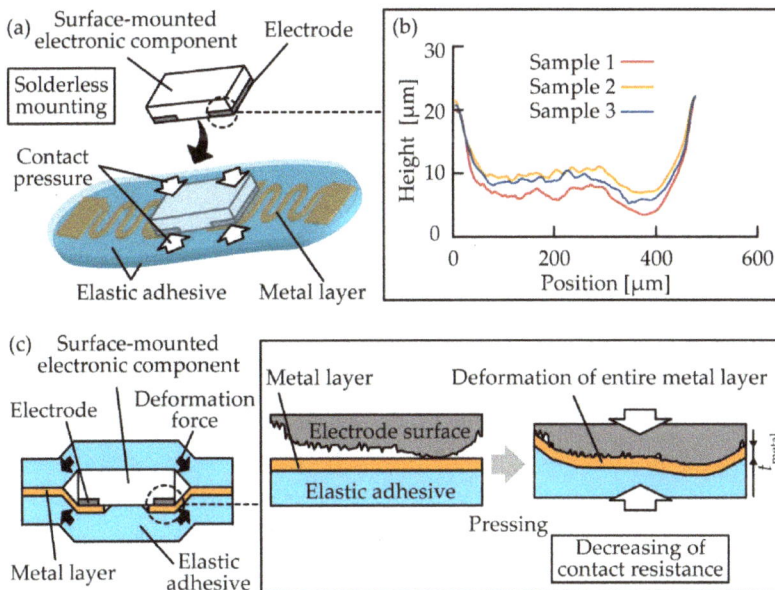

Figure 1. (**a**) Schematic image of electronic component mounting through contact pressure using an elastic adhesive; (**b**) electrode surface profiles of surface-mounted electronic component; (**c**) deformation of contact pad of metal layer of elastic adhesive sheet by gradually increasing contact pressure.

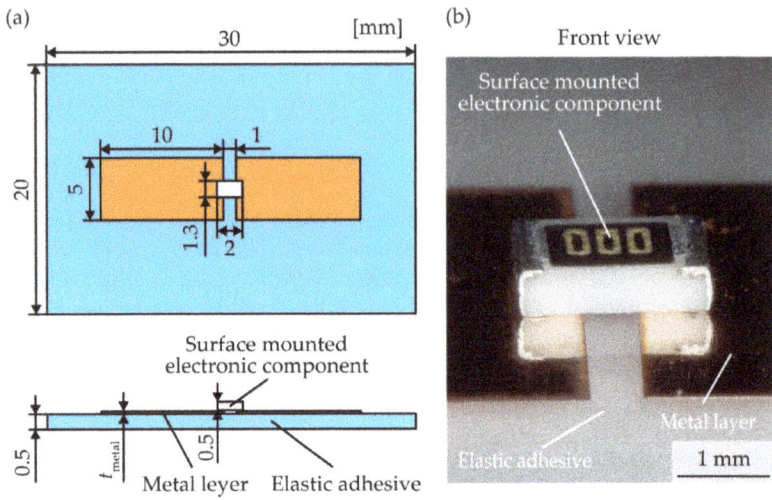

Figure 2. (**a**) Schematic images of experimental sample; (**b**) optical image of sample.

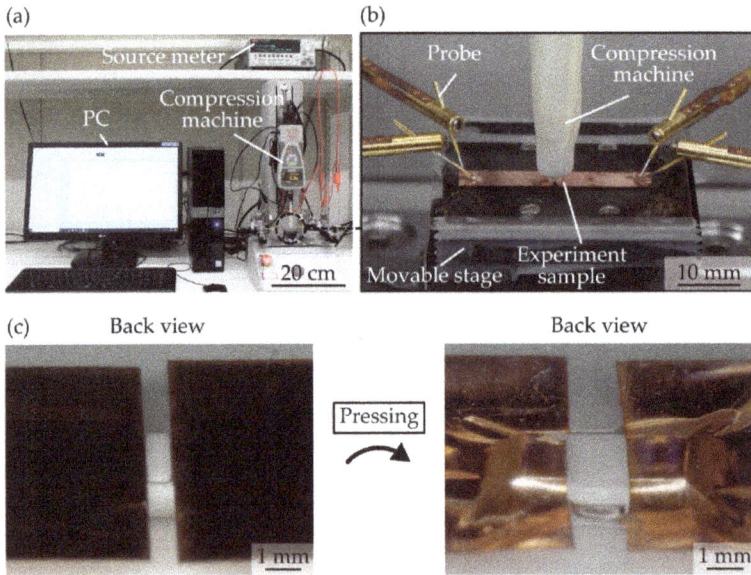

Figure 3. Optical images of experimental apparatus used to measure the change in contact resistance. (**a**) Optical image of apparatus; (**b**) optical image of enlarged view of experimental sample pressed by compression testing machine; (**c**) optical images of experimental sample with and without pushing.

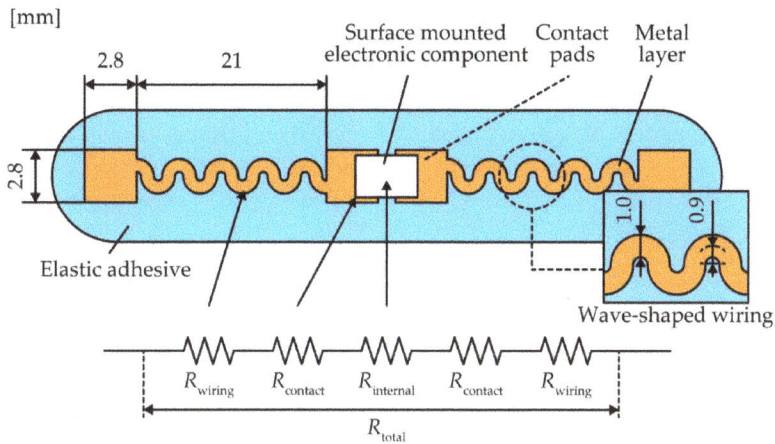

Figure 4. Schematic image of sample device.

Based on the results obtained by measuring the above parameters, we fabricated several types of electrode structures to obtain lower resistance using higher pressure and investigated the relationship between the structures and compression force. We fabricated several samples of the simple adhesive, concave, and sandwich structures and compared the contact resistivity of each structure. In this measurement, we used an elastic adhesive with a thickness of 0.5 mm and a metal layer of rolled copper films with a thickness of 5 μm for device fabrication. The depth of the concave region was 0.5 mm because we used two elastic adhesive sheets to make the concave structure. The metal layers were patterned to the contact pad with a length of 10 mm and width of 5 mm formed via laser cutting. Then, the electrodes were transferred onto the elastic adhesive with a gap of 1 mm, and a 0.5-mm-thick chip resistor (MCR10EZPJ000, ROHM Co., Kyoto, Japan) as an electrical component was placed on it. The contact resistivity ($r_{contact}$) was measured using the source meter 10 min after connecting. In addition, we measured the contact resistivity in the sandwich structure with other surface-mounted electronic components to investigate the encapsulation of the sandwich structure. A 0.6-mm-thick chip resistor (RK73ZW2HTTE, KOA CORPORATION, Tokyo, Japan), a 0.64-mm-thick chip resistor (3522ZR, TE Connectivity Ltd., Kanagawa, Japan), and a 1.1-mm-thick chip resistor (WSL251200000ZEA9, Vishay Intertechnology, Inc., Pennsylvania, USA) were mounted in a sandwich structure in the same manner as the chip resistor (MCR10EZPJ000, ROHM Co., Kyoto, Japan). Finally, to demonstrate the application of the sandwich structure for flexible electrical circuits, we fabricated an electronic device. A surface-mounted light emitting diode (LED) chip (OSR50805C1C, OptoSupply, N.T., Hong Kong, China) was embedded in the sandwich structure using a 5-μm-thick copper film and 0.5-mm-thick acrylic foam adhesive sheets to form an electrical contact.

3. Results and Discussion

Figure 5 shows the change in the contact resistivity ($r_{contact}$) against various contact pressures ($P_{contact}$) with various thicknesses of the copper contact pads (t_{metal}). $r_{contact}$ is the contact resistance per unit contact area. $r_{contact}$ was much higher than 10^{-2} Ω·m² at 0 kPa; it then dramatically decreased to 10^{-7} Ω·m² as $P_{contact}$ increased for each t_{metal}. Then, $r_{contact}$ gradually decreased to as low as approximately 10^{-8} Ω·m² after $P_{contact}$ increased. Moreover, the contact pressure values with an $r_{contact}$ of 1.0×10^{-7} Ω·m² for each t_{metal} are shown in Figure 5. The pressure value for an $r_{contact}$ of 1.0×10^{-7} Ω·m² decreased as t_{metal} decreased; these results indicate that the thin electrode decreases the pressure value at an $r_{contact}$ of 1.0×10^{-7} Ω·m². The sample device fabricated to measure and compare the total resistance, contact resistance ($R_{contact}$), and wiring resistance of metal layer (R_{wiring}) is shown in Figure 6. The copper films patterned to form contact pads and wave-shaped wirings are shown in Figure 4. $R_{contact}$ at a pressure of 800 kPa was measured as shown in Figure 3. Figure 6 shows $R_{contact}$, R_{wiring}, $R_{internal}$, and R_{total} for various values of t_{metal}. $R_{contact}$ decreased from 131.5 to 49.6 mΩ as t_{metal} decreased from 50 to 5 μm. R_{wiring} increased from 23.5 to 215 mΩ as t_{metal} decreased. R_{total} decreased as t_{metal} decreased from 50 to 30 μm, and the resistance reached a minimum value of 123 mΩ at a t_{metal} value of 30 μm. R_{total} remained the same as t_{metal} decreased from 30 to 10 μm, and then increased to 275 mΩ as t_{metal} decreased from 10 to 5 μm. Therefore, with these dimensions of the device, R_{total} reached a minimum value in the t_{metal} range 30–10 μm. This result indicates a trade-off point with respect to t_{metal} for minimizing R_{total}.

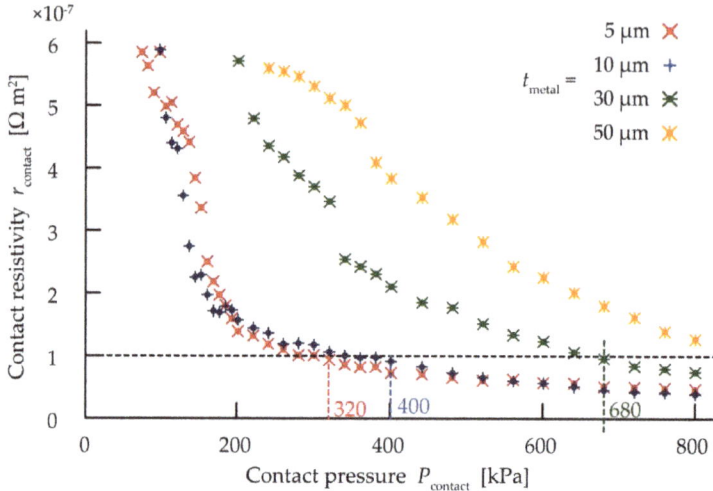

Figure 5. Relationship between contact pressure and contact resistivity with various copper layer thicknesses. The number of trials in each thickness was five.

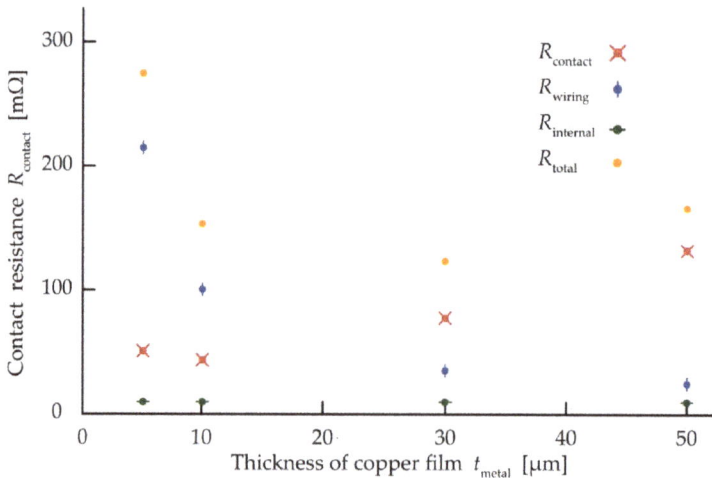

Figure 6. Relationship between thicknesses of copper layer t_{metal}, contact resistance $R_{contact}$, wiring resistance R_{wiring}, internal resistance $R_{internal}$, and total resistance R_{total}. Five trials were carried out to measure $R_{contact}$ for each thickness. Two trials were carried out to measure R_{wiring} and $R_{internal}$ for each thickness.

Figure 7 shows schematics and optical images of the structures based on three different designs with the considered contact pressure and metal layer thickness for electrical connection from the above measurement results. In this measurement, we used an elastic adhesive with a thickness of 0.5 mm, a metal layer of rolled copper films with a thickness of 5 μm, and a surface-mounted electronic component with a height of 0.5 mm for device fabrication. Figure 7a is a simple adhesive structure, in which a surface-mounted electronic component is simply placed on the contact pads of the metal layer on the elastic adhesive. In this design, first, the electronic component is pressed against the contact pads through an external force, and the electronic component contacts the elastic adhesive. Then, after removing the external force, the elastic adhesive under the electronic component is deformed, and the restoring force of the deformed elastic adhesive presses the electronic component against the contact pads. As a result, an electrical connection is caused because of the restoring force. The restoring force based on the deformation of the elastic adhesive is weak, however, and the contact resistance is expected to be high. Therefore, we proposed a concave structure and a sandwich structure to increase the restoring force and obtain lower contact resistance in different ways. Figure 7b shows schematic images and an optical image of a concave structure, which is used with the non-uniform thickness substrate of the elastic adhesive. Owing to the concave shape of the elastic adhesive under the surface-mounted electronic component, the bottom of the concave shape of the elastic adhesive is stretched more, and the electronic component is pulled down stronger than in the simple structure. Figure 7c shows schematic images and an optical image of a sandwich structure. A surface-mounted electronic component is placed on the contact pads on a base layer of the elastic adhesive; then, an upper layer containing the elastic adhesive is placed on the electronic component and pressed via an external force to contact the base layer strongly. After removing the external force, the elastic adhesives are stretched to press the electronic component against the contact pads by providing a restoring force.

Further, we fabricated testing structures and obtained contact resistivity ($r_{contact}$) values of 2.5×10^{-2}, 1.9×10^{-2}, and 8.0×10^{-8} Ω·m^2 for the simple adhesive, concave, and sandwich structures, respectively, as shown in Figure 8a. Though the contact pressure value in these structures cannot be directly measured because the contact pressure is an internal pressure, we can estimate the contact pressure value from the contact resistance value based on the relationship between the contact pressure and the contact resistance explained in Figure 5. According to the values of the contact resistance in Figure 8a, the values of the contact pressure were estimated at less than approximately 60 kPa for the simple adhesive and concave structures, and more than approximately 320 kPa for the sandwich structure. The contact resistivity in the sandwich structure was on the order of 10^6 Ω·m^2 lower than that in the simple adhesive structure. This result indicates that the contact pressure was not sufficient to obtain low contact resistivity with the simple adhesive structures or concave structure. In the case of the concave structure, though high contact pressure was expected, it was considered that the elastic adhesive under the metal layer was deformed via tilting, and the contact pressure was decreased. The high contact pressure is also considered to be able to prevent the electrical component from slipping off the contact pads of the metal layer when a flexible electronic device using contact pressure is stretched or bent. Therefore, we are considering that the contact pressure is important for not only the contact resistance but also the mechanical stability. In addition, we confirmed that the contact resistivity in electronic components with different heights between 0.5 and 1.1 mm mounted by the sandwich structure was less than 5×10^{-7} Ω·m^2. Therefore, our method can use electronic components with different heights. Figure 8b shows that an LED device fabricated using a sandwich structure can function as an electronic device. In this experiment, the light intensity of the LED device was not affected by the acrylic foam adhesive sheet because of the high transparency. We confirmed that the chip LED continued emitting light for ten hours. Because the chip LED seemed to keep almost same brightness ten hours later, we could consider that the contact resistivity of the chip LED was kept on the order of 10^{-8} Ω·m^2 for more than ten hours. These results indicate that a low contact resistance can be obtained using the proposed sandwich structure, which facilitates the development of flexible electrical circuits through a simple and low-cost process.

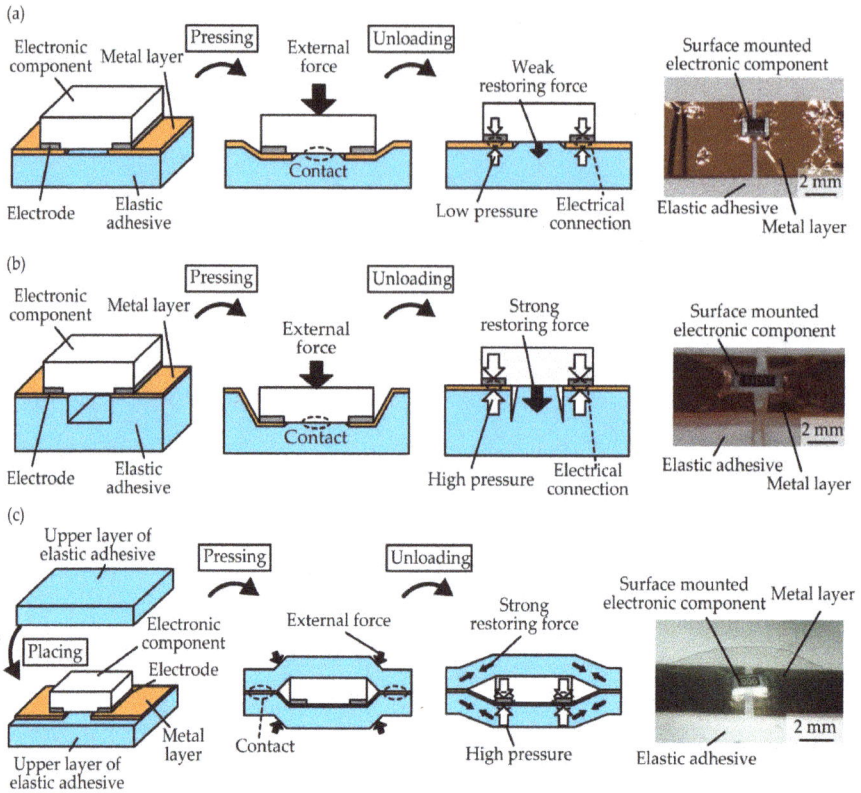

Figure 7. Schematic images and optical images of the (**a**) simple adhesive structure, (**b**) concave structure, and (**c**) sandwich structure.

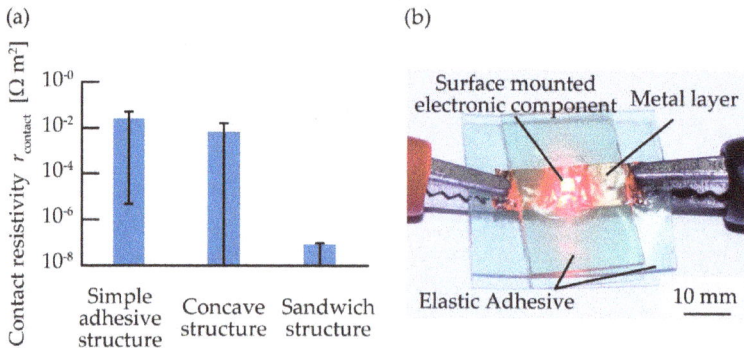

Figure 8. (**a**) Comparison of contact resistivity values in simple adhesive, concave, and sandwich structures; (**b**) flexible electronic device with light emitting diode (LED) chip mounted on a sandwich structure.

4. Conclusions

We achieved a sufficiently small contact resistivity of 8.0×10^{-8} $\Omega \cdot m^2$ between a surface-mounted electronic component and a flexible circuit substrate through pressure using an elastic adhesive.

Micromachines **2019**, *10*, 396

First, we investigated the mechanism of contact resistance change to obtain a low contact resistance. The change in contact resistance was measured at various pressures and using copper films having thicknesses of 5, 10, 30, and 50 μm. As a result, the contact resistivity decreased to below 10^{-2} $\Omega \cdot m^2$ when the pressure increased to approximately 200 kPa. Above 200 kPa, the contact resistivity gradually decreased to 10^{-8} $\Omega \cdot m^2$ as the contact pressure increased. Based on these measurement results, we designed a sandwich structure to obtain a resistivity of 8.0×10^{-8} $\Omega \cdot m^2$. Moreover, we fabricated a simple flexible electronic device with an chip LED using the sandwich structure, and the chip LED continued emitting light for ten hours after mounting.

Author Contributions: T.S., T.K., and E.I. conceived and designed the experiments; T.S. performed the experiments; T.S. and T.K. analyzed the data; T.S. wrote the paper; T.K. and E.I. reviewed and edited the paper; E.I. supervised the research.

Funding: This research was partially supported by JST CREST Grant Number JPMJCR16Q5, Japan.

Acknowledgments: The author would like to acknowledge Shunsuke Yamada at Waseda University for his support.

Conflicts of Interest: The authors declare no conflict of interest.

References

1. Abu-Khalaf, J.; Saraireh, R.; Eisa, S.; Al-Halhouli, A. Experimental characterization of inkjet-printed stretchable circuits for wearable sensor applications. *Sensors* **2018**, *18*, 3476. [CrossRef] [PubMed]

2. Choi, M.; Jang, B.; Lee, W.; Lee, S.; Kim, T.W.; Lee, H.J.; Kim, J.H.; Ahn, J.H. Stretchable Active matrix inorganic light-emitting diode display enabled by overlay-aligned roll-transfer printing. *Adv. Funct. Mater.* **2017**, *27*, 1606005. [CrossRef]

3. Yokota, T.; Zalar, P.; Kaltenbrunner, M.; Jinno, H.; Matsuhisa, N.; Kitanosako, H.; Tachibana, Y.; Yukita, W.; Koizumi, M.; Someya, T. Ultraflexible organic photonic skin. *Sci. Adv.* **2016**, *2*, e1501856. [CrossRef] [PubMed]

4. Sekitani, T.; Nakajima, H.; Maeda, H.; Fukushima, T.; Aida, T.; Hata, K.; Someya, T. Stretchable active-matrix organic light-emitting diode display using printable elastic conductors. *Nat. Mater.* **2009**, *8*, 494–499. [CrossRef] [PubMed]

5. Hu, X.; Krull, P.; De Graff, B.; Dowling, K.; Rogers, J.A.; Arora, W.J. Stretchable inorganic-semiconductor electronic systems. *Adv. Mater.* **2011**, *23*, 2933–2936. [CrossRef] [PubMed]

6. Liu, B.; Zhang, J.; Wang, X.; Chen, G.; Chen, D.; Zhou, C.; Shen, G. Hierarchical three-dimensional $ZnCo_2O_4$ nanowire arrays/carbon cloth anodes for a novel class of high-performance dlexible lithium-ion batteries. *Nano Lett.* **2012**, *12*, 3005–3011. [CrossRef]

7. Lee, J.W.; Xu, R.; Lee, S.; Jang, K.-I.; Yang, Y.; Banks, A.; Yu, K.J.; Kim, J.; Xu, S.; Ma, S.; et al. Soft, thin skin-mounted power management systems and their use in wireless thermography. *Proc. Natl. Acad. Sci. USA* **2016**, *113*, 6131–6136. [CrossRef]

8. El-Kady, M.F.; Kaner, R.B. Scalable fabrication of high-power graphene micro-supercapacitors for flexible and on-chip energy storage. *Nat. Commun.* **2013**, *4*, 1475. [CrossRef]

9. Xu, S.; Zhang, Y.; Cho, J.; Lee, J.; Huang, X.; Jia, L.; Fan, J.A.; Su, Y.; Su, J.; Zhang, H.; et al. Stretchable batteries with self-similar serpentine interconnects and integrated wireless recharging systems. *Nat. Commun.* **2013**, *4*, 1543. [CrossRef]

10. Pushparaj, V.L.; Shaijumon, M.M.; Kumar, A.; Murugesan, S.; Ci, L.; Vajtai, R.; Linhardt, R.J.; Nalamasu, O.; Ajayan, P.M. Flexible energy storage devices based on nanocomposite paper. *Proc. Natl. Acad. Sci. USA* **2007**, *104*, 13574–13577. [CrossRef]

11. Lipomi, D.J.; Vosgueritchian, M.; Tee, B.C.K.; Hellstrom, S.L.; Lee, J.A.; Fox, C.H.; Bao, Z. Skin-like pressure and strain sensors based on transparent elastic films of carbon nanotubes. *Nat. Nanotechnol.* **2011**, *6*, 788–792. [CrossRef] [PubMed]

12. Kim, J.; Lee, M.; Shim, H.J.; Ghaffari, R.; Cho, H.R.; Son, D.; Jung, Y.H.; Soh, M.; Choi, C.; Jung, S.; et al. Stretchable silicon nanoribbon electronics for skin prosthesis. *Nat. Commun.* **2014**, *5*, 1–11. [CrossRef] [PubMed]

13. Someya, T.; Sekitani, T.; Iba, S.; Kato, Y.; Kawaguchi, H.; Sakurai, T. A large-area, flexible pressure sensor matrix with organic field-effect transistors for artificial skin applications. *Proc. Natl. Acad. Sci. USA* **2004**, *101*, 9966–9970. [CrossRef] [PubMed]

14. Yokota, T.; Inoue, Y.; Terakawa, Y.; Reeder, J.; Kaltenbrunner, M.; Ware, T.; Yang, K.; Mabuchi, K.; Murakawa, T.; Sekino, M.; et al. Ultraflexible, large-area, physiological temperature sensors for multipoint measurements. *Proc. Natl. Acad. Sci. USA* **2015**, *112*, 14533–14538. [CrossRef] [PubMed]

15. Shih, W.P.; Tsao, L.C.; Lee, C.W.; Cheng, M.Y.; Chang, C.; Yang, Y.J.; Fan, K.C. Flexible temperature sensor array based on a Graphite-Polydimethylsiloxane composite. *Sensors* **2010**, *10*, 3597–3610. [CrossRef] [PubMed]

16. Rogers, J.A.; Someya, T.; Huang, Y.; Sorensen, A.E.; Lian, J.; Greer, J.R.; Valdevit, L.; Carter, W.B.; Ge, Q.; Jackson, J.A.; et al. Materials and mechanics for stretchable electronics. *Science* **2010**, *327*, 1603–1607. [CrossRef] [PubMed]

17. Song, J. Mechanics of stretchable electronics. *Curr. Opin. Solid State Mater. Sci.* **2015**, *19*, 160–170. [CrossRef]

18. Hammock, M.L.; Chortos, A.; Tee, B.C.K.; Tok, J.B.H.; Bao, Z. 25th anniversary article: The evolution of electronic skin (E-Skin): A brief history, design considerations, and recent progress. *Adv. Mater.* **2013**, *25*, 5997–6038. [CrossRef] [PubMed]

19. Trung, T.Q.; Lee, N.E. Materials and devices for transparent stretchable electronics. *J. Mater. Chem. C* **2017**, *5*, 2202–2222. [CrossRef]

20. Dimitrakopoulos, C.D.; Malenfant, P.R.L. Organic thin film transistors for large area electronics. *Adv. Mater.* **2002**, *14*, 99–117. [CrossRef]

21. Zhou, Y.; Fuentes-Hernandez, C.; Shim, J.; Meyer, J.; Giordano, A.J.; Li, H.; Winget, P.; Papadopoulos, T.; Cheun, H.; Kim, J.; et al. A universal method to produce low–work function electrodes for organic electronics. *Science* **2012**, *336*, 327–332. [CrossRef] [PubMed]

22. Liao, C.; Zhang, M.; Yao, M.Y.; Hua, T.; Li, L.; Yan, F. Flexible organic electronics in biology: Materials and devices. *Adv. Mater.* **2015**, *27*, 7493–7527. [CrossRef] [PubMed]

23. Harrison, M.R.; Vincent, J.H.; Steen, H.A.H. Lead-free reflow soldering for electronics assembly. *Solder. Surf. Mt. Technol.* **2001**, *13*, 21–38. [CrossRef]

24. Zhou, J.; Sun, Y.; Xue, F. Properties of low melting point Sn-Zn-Bi solders. *J. Alloys Compd.* **2005**, *397*, 260–264. [CrossRef]

25. Mir, I.; Kumar, D. Recent advances in isotropic conductive adhesives for electronics packaging applications. *Int. J. Adhes. Adhes.* **2008**, *28*, 362–371. [CrossRef]

26. Li, Y.; Wong, C.P. Recent advances of conductive adhesives as a lead-free alternative in electronic packaging: Materials, processing, reliability and applications. *Mater. Sci. Eng. R* **2006**, *51*, 1–35. [CrossRef]

27. Okamoto, M.; Kurotobi, M.; Takeoka, S.; Sugano, J.; Iwase, E.; Iwata, H.; Fujie, T. Sandwich fixation of electronic elements using free-standing elastomeric nanosheets for low-temperature device processes. *J. Mater. Chem. C* **2017**, *5*, 1321–1327. [CrossRef]

28. Mitsui, R.; Takahashi, S.; Nakajima, S.I.; Nomura, K.I.; Ushijima, H. Film-type connection system toward flexible electronics. *Jpn. J. Appl. Phys.* **2014**, *53*, 05HB04. [CrossRef]

micromachines

MDPI

Article

Connecting Mechanism for Artificial Blood Vessels with High Biocompatibility

Ai Watanabe and Norihisa Miki *

Department of Mechanical Engineering, Keio university, 3-14-1 Hiyoshi, Kohoku-ku, Yokohama,
Kanagawa 223-8522, Japan
* Correspondence: miki@mech.keio.ac.jp; Tel.: +81-45-566-1430

Received: 27 May 2019; Accepted: 25 June 2019; Published: 28 June 2019

Abstract: This paper proposes a connecting mechanism for artificial vessels, which can be attached/detached with ease and does not deteriorate the biocompatibility of the vessels at the joint. The inner surface of the artificial vessels is designed to have high biocompatibility. In order to make the best of the property, the proposed connecting mechanism contacts and fixes the two artificial vessels whose contacting edges are turned inside out. In this manner, blood flowing inside the vessels only has contact with the biocompatible surface. The biocompatibility, or biofouling at the joint was investigated after in vitro blood circulation tests for 72 h with scanning electron microscopy. Blood coagulation for a short term (120 min) was evaluated by activated partial thromboplastin time (APTT). A decrease of APTT was observed, although it was too small to conclude that the connector augmented the blood coagulation. The micro dialysis device which our group has developed as the artificial kidney was inserted into the blood circulation system with the connector. Decrease of APTT was similarly small. These experiments verified that the proposed connector can be readily applicable for implantable medical devices.

Keywords: connector; artificial blood vessel; medical device; blood coagulation; implant; artificial kidney; biocompatible

---

1. Introduction

Micro/nano technologies have enabled miniaturized medical devices [1–5]. When they are small enough to be implanted, they can monitor and/or treat the patients continuously without bothering their daily lives. Our group has been developing micro filtering devices which we aim to use as an artificial kidney [6–8]. Currently, there are over 320,000 hemodialysis patients in Japan and 2.6 million patients in the world [9,10]. The hemodialysis therapy is well developed, particularly in Japan, however it leads to low quality of life of the patients. The patients are mandated to visit hospital three times a week where they receive the treatment for 4 h. Frequent punctures do not only give pain to the patients but also damage the blood vessels. The patients have severe restriction in water and salt intake. Implantable artificial kidneys will alleviate these problems and drastically improve the quality of life of patients.

One of the major challenges of the artificial kidney is biofouling, which deteriorates the dialysis performance and may mandate the device to be replaced. In order to simplify the surgery procedures and alleviate invasiveness to the patients, we propose to use a connector, as shown in Figure 1. The device is connected to the blood vessels with the biocompatible artificial vessels, which can be separated at the connector. In replacement of the device, or maintenance surgery, the access to the blood vessels is preserved and only the device is exchanged. When the device is implanted at the shallow region beneath the skin, the surgery will be of further ease.

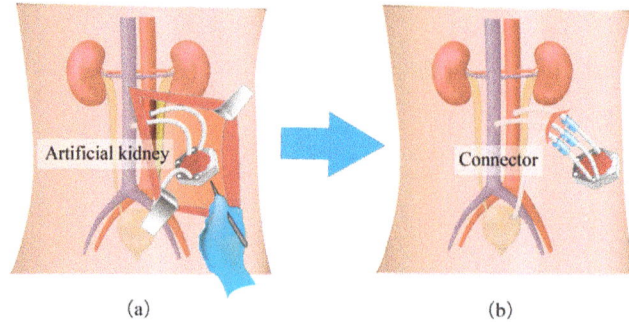

Figure 1. Maintenance surgery of the artificial kidney (**a**) without and (**b**) with the connecting system. It allows the device to be replaced while the connections between the artificial vessels and the artery, vein, and bladder are maintained. The surgery can be less invasive and easier by implanting the device at the shallow region beneath the skin.

The requirements for the connector include high biocompatibility, as well as ease of manipulation. In this paper, we propose a connector mechanism which makes the best of the high biocompatibility of the artificial vessels [11–13]. The connector brings the artificial vessels in contact, with their edges being turned inside out, so that blood only comes in contact with the highly biocompatible inner surface of the artificial vessels. It has a snap-fit mechanism for simple and firm connection. The detailed mechanism was discussed and its biocompatibility, or blood coagulability, of the connector was experimentally assessed. The proposed connector mechanism can be readily applicable to other implantable medical devices.

2. Design and Assembly of the Connector

2.1. Biocompatibility

The biocompatibility that we focus on in this work is blood anti-coagulation. Blood coagulation leads to the formation of blood clots and the clogging of the medical device. In case of the artificial kidney, its filtration capacity deteriorates with the coagulation. One of the major factors of blood coagulation is the contact of blood to foreign body surfaces [14]. Another factor is turbulence in the blood flow which can be caused by the geometry of the flow paths [15–17]. The proposed connector can be the solution for both factors. The blood anti-coagulation can be assessed by optical investigation of thrombus at the surface and activated partial thromboplastin time (APTT), which represents the speed of blood coagulation [17,18].

2.2. Mechanism Design and Assembly

In order to exploit the high biocompatibility of the artificial blood vessels, the connector is designed such that blood only contacts the inner surfaces of the artificial blood vessels to be connected.

The connector consists of the cylindrical parts A and B made of metal and the snap-fit mechanism that was manufactured by 3D printing, as shown in Figure 2a. The artificial vessels are the ePTF graft (thin-wall straight type, W. L. Gore & Associates, Co., Ltd., Tokyo, Japan). The inner and outer diameter are 6.0 mm and 6.8 mm, respectively. The parts A and B are manufactured from stainless steel (SUS304) with inner and outer diameters of 6.8 mm and 7.6 mm and 8.2 mm and 9.0 mm, respectively. The outermost parts for the snap-fit mechanism are 3D-printed from Nylon 12.

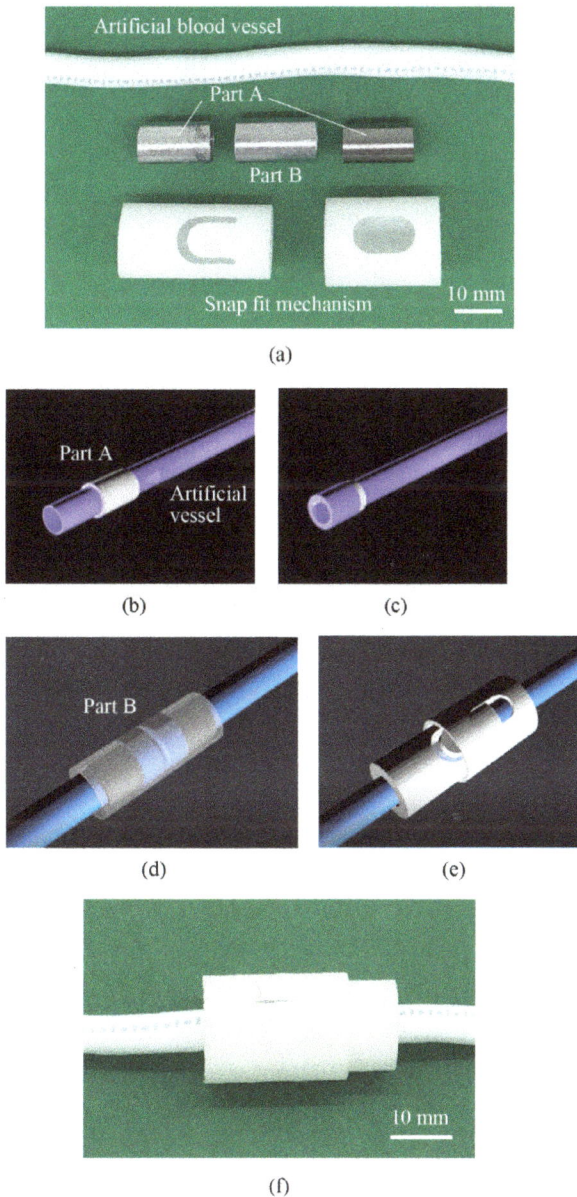

Figure 2. Proposed connecting mechanism of artificial blood vessels. (**a**) The mechanism consists of metal part A and B, 3D-printed snap fit mechanism, and artificial vessels. (**b,c**) The edges of the vessels to be joined are turned inside out over part A. (**d**) They are brought into contact inside part B. (**e**) The snap fit mechanism secures the connection. (**f**) The photo of the connected mechanism. The resulting size of the connection mechanism is 15.4 mm in diameter and 29.0 mm in length.

First, the artificial vessels are inserted into the cylindrical part A and the edges to be contacted are turned inside out (Figure 2b,c shows the case for one artificial vessel). They are brought into contact and the cylindrical part B covers the joint part, as shown in Figure 2d. The 3D-printed snap-fit mechanism is brought over part B and locks the connection (Figure 2e). Figure 2f shows the photo of the joint part with the connector. The assembly completes in 1 min, which we consider sufficiently short. The fixing with the case was experimentally verified to be strong enough; no detachment or leak was observed during the experiments. At the joint, a small ditch at the wall surface will be formed, as shown in Figure 3. Biocompatibility at the ditch will be investigated later.

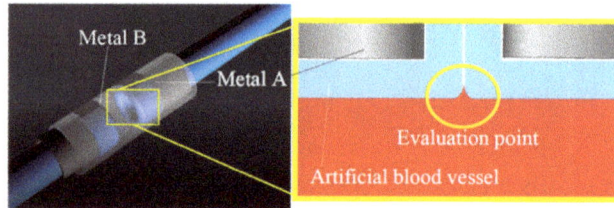

Figure 3. Joint area inside the connector. A small ditch is formed at the interface, where blood coagulation might be promoted.

Since the mechanism is designed such that the blood only flows through the artificial vessels without any deformation, the pressure drop across the connector was negligible, which was later verified in the experiments.

2.3. Micro Filtering Device as the Artificial Kidney

The micro filtering device consists of nano porous polyether sulfone (PES) membranes and microfluidic channels made of Titanium (The Nilaco Corporation, Tokyo, Japan). PES membranes are formed by the wet inversion method from PES (Sumitomo Chemical Grade 4800P, Sumitomo Chemical, Co., Ltd., Tokyo, Japan), poly(ethylene glycol) (PEG; molecular weight of 1000, Wako Pure Chemical Industries, Ltd., Osaka, Japan), and N,N-dimethylacetamide (DMAc; Wako Pure Chemical Industries, Ltd.). The microfluidic channels are formed by electrolytic etching. The PES membranes and microfluidic layers are stacked in sequence. The details of the device and the fabrication processes are described elsewhere [8,19,20].

3. Experimental Methods

Experiments were conducted based on ISO 10993, the biological evaluation about the medical machine. Thrombus formation and blood coagulation was investigated [21–23].

3.1. Thrombus Formation Tests In Vitro

Figure 4a–c illustrates the blood circulation system that includes (a) the connector, (b) two connectors, and (c) the connector and the device, respectively. Thrombus formation at the connection part was investigated in the blood circulation system with one connector (Figure 4a). Human whole blood type A (KOJ, Cosmo Bio, Japan) was used for the experiments. In practical use, the blood flow rate through the connector is expected to be several tens of mL/min. In our prior work, thrombus formation took place in the area where the blood flow rate was low [8]. In this work, since we would like to highlight the thrombus formation and change of blood coagulability, we set the blood flow rate to be 1 mL/min. In case of the setup shown in Figure 4a, the resulting pressure was 90–100 mmHg, which was measured and recorded by polygraph. The fluctuation of the pressure was induced by the pulsation of the pump. The blood contains 3.2% sodium citrate as anti-coagulate and exchanged every 24 h. After 1, 24, 48, and 72 h of circulation, the surface of the artificial blood vessels near the ditch part was treated for cell fixation and then optically investigated with scanning electron microscopy. The cell fixation process included a rinse with PBS, cell fixation with glutaraldehyde solution (2.5%

glutaraldehyde, 50% PBS, 47.5% DI water) for 1 d followed by 100% glutaraldehyde for 1 h, rinse with ethanol, and osmium coating for the SEM inspection.

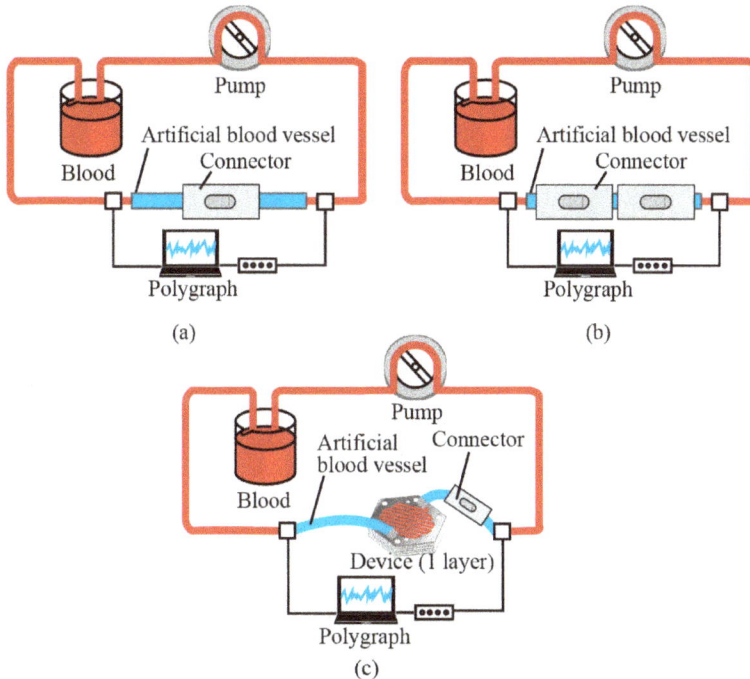

Figure 4. Blood circulation circuit with (**a**) one connector, (**b**) two connectors and (**c**) the connector and the fluidic device. The polygraph measures the pressure before and after the connectors and the device. Blood coagulation at the contact point and change of the activated partial thromboplastin time (APTT) was investigated.

3.2. Blood Coagulation Tests In Vitro

Activated partial thromboplastin time (APTT) was measured to assess the blood coagulability caused by the connector. Blood circulation circuits without and with two connectors (Figure 4b) were tested. The flow rate was set to be 1 mL/min and the pressure was 90–120 mmHg. APTT was measured at the beginning and then every 20 min until 120 min.

The APTT measurement protocol is centrifugal extraction of plasma (200 rpm, 11 min), incubation with APTT test reagents for 180 s, and measurement of APTT with calcium chloride solution.

Since the blood contacts air and the quality degrades in 120 min, the in vitro experiments cannot be continued longer than 120 min. In the previous experiment to investigate the thrombus formation, we exchanged the blood, which is not suitable for this experiment. Long term in vivo experiments will be conducted to further validate the effect of the connectors.

3.3. Blood Coagulation Tests with the Micro Filtering Device In Vitro

Effect of the micro filtering on the blood coagulation was experimentally investigated. Blood circulation systems including the micro filtering device and both the connector and the micro filtering device (Figure 4c) were prepared. In the experiments, the one-layer filtering device was used. At the beginning, the device is filled with a solution of heparin Na (Mochida Pharmaceutical, Japan) and physiological saline (Otsuka Pharmaceutical, Japan) with a ratio of 1:9 in order to prevent the initial adhesion and coagulation of proteins inside the device. APTT was measured at the beginning and then every 20 min until 120 min (7 data points). APTT measurement protocol is described in Section 3.2. The flow rate was set to be 1 mL/min and the resulting pressure was 90–120 mmHg.

4. Experimental Results and Discussion

4.1. Thrombus Formation inside the Artificial Blood Vessel

Figure 5 shows the SEM images of the artificial blood vessel in which blood flowed for 1, 24, 48, 72 h. Aggregates of proteins approximately 50 μm in size were found on the surface after 1 h. After 24 h the number of the aggregates increased and they were found inside the fiber structures. Large aggregates on the order of 1 mm in size were found after 48 h. After 72 h, these large aggregates were still found though the number of them did not increase significantly.

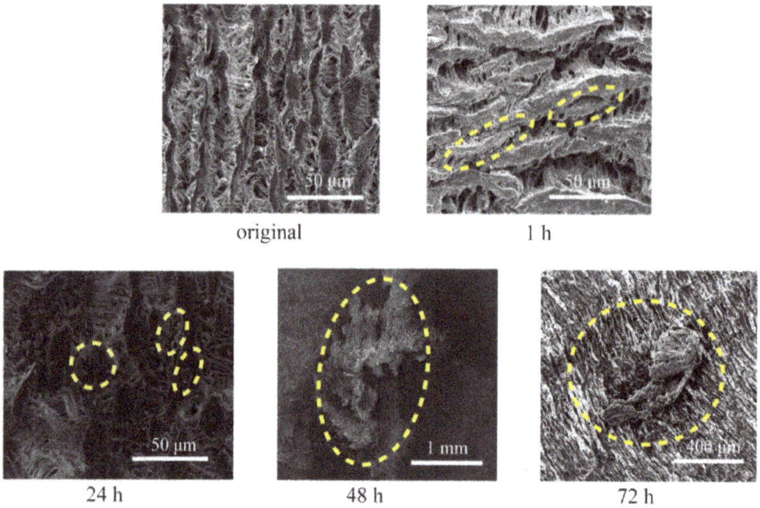

Figure 5. Thrombus formation near the joint observed by SEM. Thrombus formation was observed at each interval. However, it was not concluded that the ditch at the interface initiated the formation.

The ditch at the interface of the two artificial blood vessels was suspected to initiate blood coagulation. However, in this experiment the thrombus formation was not limited to the ditch part and no significant effect was found at least for 72 h. It was reported that the thrombosis film was formed on the surface of the artificial vessels, on which vascular endothelial cells subsequently settled. This augments the long-term stability of the artificial vessels [24,25]. We expect the artificial vessels connected with the proposed connector would exploit this phenomenon, though longer-term experiments in vivo will be necessary.

4.2. Blood Coagulation Caused by the Connector

Figure 6 shows the change of APTT with each time interval with or without the connectors. The typical APTT of human blood without anticoagulant is 24–34 s. The APTT was extended to 57–60 s with 3.2% sodium citrate. APTT was found to gradually decrease with time. No significant effect of the connectors was observed for 120 min. This indicates that the proposed connectors do not promote blood coagulation.

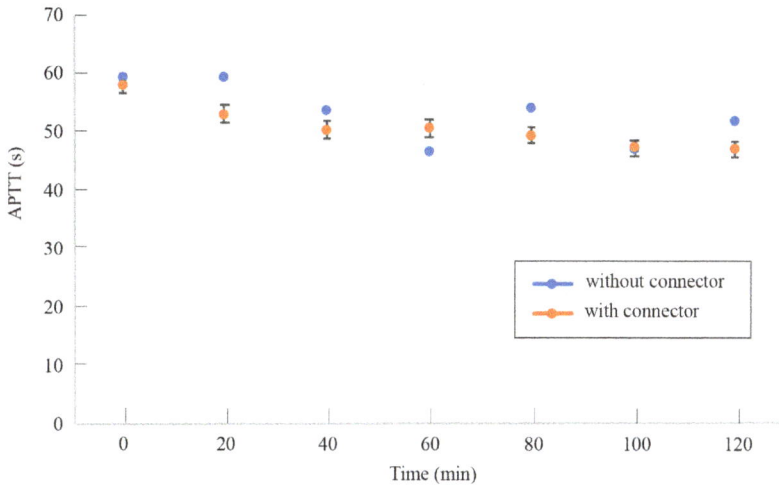

Figure 6. Progress of APTT with and without the connector. No significant differences were induced by the connector during 120 min.

Since the quality of blood degrades in 120 min, the in vitro experiments cannot be continued longer than 120 min. Long term in vivo experiments will be conducted to further validate the effect of the connectors.

4.3. Blood Coagulation Caused by the Connector and the Filtering Device

Figure 7 shows the variation of APTT with time. For all three cases, i.e., no additional components, with the one-layer micro filtering device, and with the device and the connector, APTT was found to be within the range of 50–60 s.

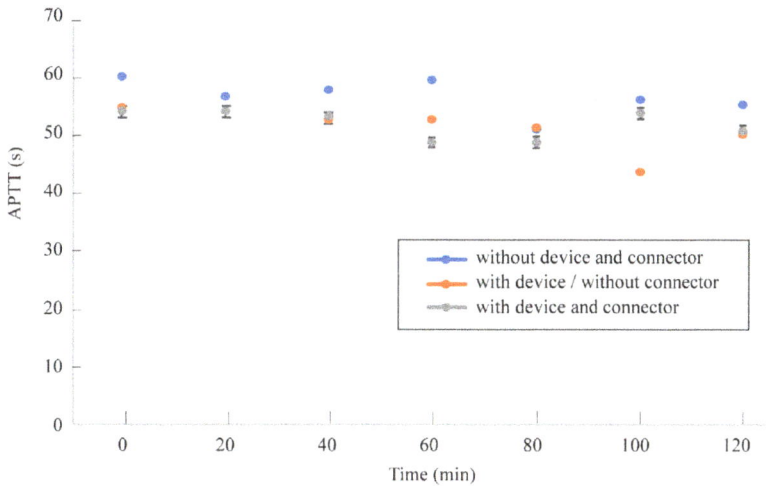

Figure 7. Progress of APTT with the device and the connector. During 120 min of experiments, no significant differences were observed.

APTT measurement for 120 min did not show any significant effects of the connectors. The longer experiments need to be conducted to verify the long-term stability of blood coagulability. However, in the in vitro experiments, the property of the blood changes with time and the resulting APTT does not reflect the effect of the connectors. Therefore, long-term in vivo experiments need to be conducted, where control of the anticoagulant concentration in blood will be challenging.

5. Conclusions

We designed and demonstrated the enfold connecting system of artificial blood vessels. The design allows blood to contact only the highly biocompatible surfaces of the artificial vessels. Optical investigation after 72 h of blood circulation did not show any significant differences between the joint part (ditch) and the other parts. APTT measurement for 120 min verified that the connector did not augment blood coagulability. The connector proposed herein can be readily applicable to simplify the surgical process of implantable medical devices without degrading the biocompatibility.

Author Contributions: Conceptualization, A.W. and N.K.; data curation, A.W. and N.K.; formal analysis, N.K.; investigation, N.K.; methodology, A.W.; project administration, N.K.; resources, N.K.; supervision, N.K.; Validation, A.W.; writing—original draft, A.W. and N.K.; writing—review and editing, N.K.

Funding: This work was supported by Ishii/Ishibasi Funds and Keio Gijyuku Academic Development Funds.

Conflicts of Interest: The authors declare no conflicts of interest.

References

1. Receveur, R.A.M.; Lindemans, F.W.; de Rooij, N.F. Microsystem technollgies for implantable applications. *J. Micromech. Microeng.* **2007**, *17*, R50–R80. [CrossRef]
2. Alici, G. Towards soft robotic devices for site-specific drug delievery. *Expert Rev. Med. Devices* **2015**, *12*, 703–715. [CrossRef] [PubMed]
3. Meng, E.; Sheybani, R. Insight: Implantable mediclal devices. *Lab Chip* **2014**, *14*, 3233. [CrossRef] [PubMed]
4. Zahn, J.D.; Talbot, N.H.; Liepmann, D.; Pisano, A.P. Microfabricated polysilicon microneedles for minimally invasive biomedical devices. *Biomed. Dev.* **2000**, *2*, 295–303. [CrossRef]
5. Arai, M.; Kudo, Y.; Miki, N. Polymer-based candle-shaped microneedle electrodes for electroencephalography on hairy skin. *Jpn. J. Appl. Phys.* **2016**, *55*, 06GP16. [CrossRef]
6. Miki, N.; Kanno, Y. Development of a nanotechnology-based dialysis device. In *Home Dialysis in Japan*; Karger Publishers: Basel, Switzerland, 2012; pp. 178–183.
7. To, N.; Sanada, I.; Ito, H.; Prihandana, G.S.; Morita, S.; Kanno, Y.; Miki, N. Water-permeable dialysis membranes for multi-layered microdialysis system. *Front. Bioeng. Biotechnol.* **2015**, *3*, 70. [CrossRef]
8. Ota, T.; To, N.; Kanno, Y.; Miki, N. Evaluation of biofouling in stainless microfluidic channels for implantable multilayered dialysis device. *Jpn. J. Appl. Phys.* **2017**, *56*, 06GN10. [CrossRef]
9. Mineshima, M.; Kawanishi, H.; Ase, T.; Kwasaki, T.; Tomo, T.; Nakamoto, H. 2016 update Japanese Society for Dialysis Therapy Standard of fluids for hemodialysis and related therapy. *Ren. Replace Ther.* **2018**, *4*, 15. [CrossRef]
10. Van der Tol, A.; Lameire, N.; Morton, R.L.; Biesen, W.V.; Vanholder, R. An international analysis of dialysis services reimbursement. *Clin. J. Am. Soc. Nephrol.* **2019**, *14*, 84–93. [CrossRef]
11. Kakisis, J.D.; Liapis, C.D.; Breuer, C.; Sumpio, B.E. Artificial blood vessel: The Holy Grail of peripheral vascular surgery. *J. Vasc. Surg.* **2005**, *41*, 349–354. [CrossRef]
12. Byrom, M.J.; Ng, M.K.C.; Bannon, P.G. Biomechanics and biocompatibility of the perfect conduit—Can we build one? *Ann. Cardiothorac. Surg.* **2013**, *2*, 435–443. [CrossRef] [PubMed]
13. Ito, E.; Okano, T. Artificial blood vessels: Structure and property of blood contacting surface. *J. Surf. Finish. Soc. Jpn.* **1998**, *49*, 715–721. [CrossRef]
14. Palta, S.; Saroa, R.; Palta, A. Overview of the coagulation system. *Indian J. Anaesth.* **2014**, *58*, 515–523. [CrossRef] [PubMed]
15. Fogelson, A.L.; Guy, R.D. Platelet-wall interactions in continuum models of platelet thrombosis: Formulation and numerical solution. *Math. Med. Biol.* **2004**, *21*, 293–334. [CrossRef] [PubMed]
16. Basmadjian, D. The effect of flow and mass-transport in thrombogenesis. *Ann. Biomed. Eng.* **1990**, *18*, 685–709. [CrossRef] [PubMed]

17. Shinoda, T.; Arakura, H.; Katakura, M.; Shirota, T.; Nakagawa, S. Usefulness of thrombelastgraphy for dosage monitoring of low molecular weight heparin and unfractionated heparin during hemodialysis. *Artif. Organs* **1990**, *14*, 413–415. [CrossRef] [PubMed]

18. Chen, C.C.; You, J.Y.; Ho, C.H. The aPTT assay as a monitor of heparin anticoagulation efficacy in clinical settings. *Adv. Ther.* **2003**, *20*, 231–236. [CrossRef]

19. Prihandana, G.S.; Sanada, I.; Ito, H.; Noborisaka, M.; Kanno, Y.; Suzuki, T.; Miki, N. Antithrombogenicity of fluorinated diamond-like carbon films coated nano porous polyethersulfone (PES) membrane. *Materials* **2013**, *6*, 4309–4323. [CrossRef]

20. Ye, G.; Miki, N. Multilayered microfilter using PES nano porous membrane applicable as the dialyzer of a wearable artificial kidney. *J. Micromech. Microeng.* **2009**, *19*, 065031.

21. Weber, M.; Steinle, H.; Golombek, S.; Hann, L.; Schlensak, C.; Wendel, H.P.; Avci-Adali, M. Blood-contacting biomaterials: In vitro evaluation of the hemocompatibility. *Front. Bioeng. Biotechnol.* **2018**, *6*, 99. [CrossRef]

22. Van Oeveren, W.; Haan, J.; Lagerman, P.; Schoen, P. Comparison of coagulation activity tests in vitro for selected biomaterials. *Artif. Organs* **2002**, *26*, 506–511. [CrossRef] [PubMed]

23. Stang, K.; Krajewski, S.; Neumann, B.; Kurz, J.; Post, M.; Stoppelkamp, S.; Fennrich, S.; Avci-Adali, M.; Armbruster, D.; Schlensak, C.; et al. Hemocompatibility testing according to ISO 10993-4: Discrimination between pyrogen-and device-induced hemostatic activation. *Mater. Sci. Eng. C* **2014**, *42*, 422–428. [CrossRef] [PubMed]

24. Noishiki, N. A concept of tissue engineering in the development of small diameter vascular prothesis. *Artif. Organ.* **1998**, *27*, 601–607. [CrossRef]

25. Xue, L.; Greisler, H.P. Biomaterials in the development and future of vascular grafts. *J. Vasc. Surg.* **2003**, *37*, 472–480. [CrossRef] [PubMed]

micromachines

MDPI

Article

Extension of the Measurable Wavelength Range for a Near-Infrared Spectrometer Using a Plasmonic Au Grating on a Si Substrate

Yu Suido [1], Yosuke Yamamoto [1], Gaulier Thomas [2], Yoshiharu Ajiki [1,3] and Tetsuo Kan [1,*

[1]   Department of Mechanical Engineering and Intelligent Systems, Graduate School of Informatics and Engineering, The University of Electro-Communications, 1-5-1 Chofugaoka, Chofu-city, Tokyo 182-8585, Japan; suido@ms.mi.uec.ac.jp (Y.S.); yamamoto@ms.mi.uec.ac.jp (Y.Y.); yoshiharu_ajiki@ot.olympus.co.jp (Y.A.)

[2]   École Nationale Supérieure de Mécanique et des Microtechniques, 26 Rue de l'Épitaphe, 25000 Besançon, France; thomas.gaulier@ens2m.org

[3]   Mobile System Development Division, Imager & Analog LSI Technology Department, Olympus Corporation, 2-3 Kuboyama-cho, Hachioji-city, Tokyo 192-8512, Japan

*   Correspondence: tetsuokan@uec.ac.jp; Tel.: +81-42-443-5423

Received: 14 May 2019; Accepted: 13 June 2019; Published: 17 June 2019

Abstract: In this paper, we proposed near-infrared spectroscopy based on a Si photodetector equipped with a gold grating and extended the measurable wavelength range to cover 1200–1600 nm by improving a spectrum derivation procedure. In the spectrum derivation, photocurrent data during alteration of the incidence angle of the measured light were converted using a responsivity matrix R, which determines the spectroscopic characteristics of the photodetector device. A generalized inverse matrix of R was used to obtain the spectrum and to fit a situation where multiple surface plasmon resonance (SPR) peaks appeared in the scanning range. When light composed of two wavelengths, 1250 nm and 1450 nm, was irradiated, the two wavelengths were distinctively discriminated using the improved method.

Keywords: near-infrared; spectroscopy; surface plasmon resonance; Schottky barrier; grating; Si

1. Introduction

Near-infrared (NIR) spectroscopy is a method of analyzing an object by its optical spectrum, and this method is used in many fields, such as agriculture, chemistry, and medicine [1–4]. In recent years, compact near-infrared spectrometers have been intensively studied. Particularly, plasmonic-based spectrometers are attracting attention because dispersal of the incident light is performed with a thin layer to allow a compact optical system [5–9]. One such example is based on a plasmonic color filter where the transmission wavelength can be selected by designing periodic structures of a plasmonic metal [6–8]. The transmitted light is usually measured by a photodetector located below the plasmonic filter. In addition, an NIR spectrometer using a plasmonic grating on silicon is reported [9]. Since a Schottky barrier is formed at an interface of the grating metal/silicon (Si), coupled surface plasmon resonance (SPR) on the grating is directly detected as a current on the device. The photodetector is integrated with a plasmonic grating, thus this method is well suited for constructing compact NIR spectrometers. However, there is an issue with the Schottky-type spectrometer: The operating wavelength range is only as small as 100 nm. It is therefore required to widen the measurable wavelength range to cover the whole NIR wavelength range for this spectrometer to become practical.

In this paper, we increased the NIR spectroscopy range from 1200 nm to 1600 nm based on a Si photodetector equipped with a gold grating by improving the spectrum derivation procedure.

Although a basic spectrum derivation procedure is provided in a previous reference [9], the simple application of this procedure for a wider NIR range resulted in artifacts in the calculated results due to the multiple SPR coupling points that appeared in the angular photocurrent spectrum. We thus revised the spectrum derivation method to be applicable for a wider NIR wavelength range. We fabricated a plasmonic grating Si photodetector and integrated it with a measurement circuit for low signal noise measurement. The electrical characteristics and SPR photodetection performance of the device were investigated. The spectroscopic performance was evaluated by irradiating the device with light composed of two NIR wavelengths, and the spectroscopic results are shown in the figures below.

2. Principles of Light Detection Using Surface Plasmon Resonance (SPR)

The structure of the proposed spectrometer is shown in Figure 1a, which was almost the same as that in a previous report [9]. The device consisted of an n-type silicon substrate and a one-dimensional diffraction grating of a thin gold film formed on the n-type silicon substrate. The grooves of the diffraction grating running from the front to the back side of the paper surface are shown in Figure 1a. The gold film also served as the anode electrode. The surrounding medium around the gold grating was assumed to be air. An aluminum film was formed at the bottom surface of the device and served as a cathode electrode.

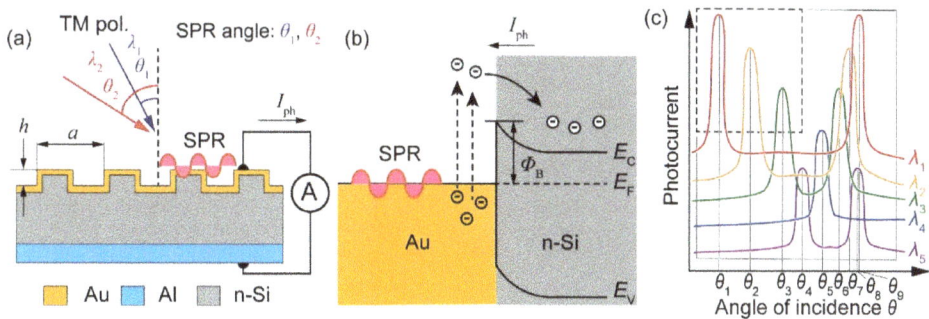

Figure 1. Proposed spectrometer configuration. (**a**) A structure of the photodetector device. (**b**) A current detection mechanism of incident light using surface plasmon resonance (SPR). (**c**) Angular photocurrent spectra for four different wavelength lights.

SPR was excited on the gold grating surface by the light incidence on the grating surface at the resonant incidence angle θ, which was dependent on the wavelength λ of the light. The SPR coupling behavior was as follows: When transverse magnetic (TM) light with wavelength λ entered the grating surface with an incidence angle θ, the incident light was diffracted by the grating. At a certain angle of incidence θ, the following SPR condition was satisfied, and SPR can be excited on the gold grating surface [10],

$$\frac{\omega}{c}\sqrt{\varepsilon_{air}}\sin\theta + \frac{2m\pi}{a} = \frac{\omega}{c}\sqrt{\frac{\varepsilon_{air}\varepsilon_{Au}}{\varepsilon_{air}+\varepsilon_{Au}}},\tag{1}$$

where ω is the angular frequency of the incident light, c is the speed of light in a vacuum, m is the order of diffraction, and ε_{air} and ε_{Au} are the relative permittivity of air and gold, respectively. The left side of Equation (1) is the wavenumber component along the device surface of the diffracted light. The right side is the wavenumber of SPR on the gold/air interface. SPR was excited on the gold surface when the angular frequency and wavenumber of the incident light and SPR coincided with each other due to the effect of diffraction. Based on this relationship, it can be confirmed that SPR occurred at different angles θ if the wavelength λ was different. In addition, since the gold grating was formed on the n-type silicon, a Schottky barrier was formed at the interface between gold and n-type silicon [11–14]. Normally, a Schottky barrier with a height Φ_B of 0.7–0.8 eV (detection limit wavelength approximately

1.55–1.77 µm) was formed there (Figure 1b) [15]. Free electrons excited by the SPR overcame the barrier and flowed from the gold grating to the n-type silicon, and the current I_{ph} had a peak value there.

Figure 1c shows schematic photocurrent I_{ph} plots with respect to the angle of incidence θ for five different wavelength lights, from λ_1 through λ_5. In this plot, the device was tilted to alter the angle of incidence θ. The photocurrent peaks corresponded to situations where Equation (1) was satisfied. The peak's angular position thus shifted when the wavelength was changed. Assuming that $\lambda_1 < \lambda_2 < \lambda_3$, the angular positions monotonically shifted from θ_1 through θ_2 to θ_3, as depicted within a dotted line square in Figure 1c. This angular shift direction, plus or minus, is determined by a sign of the corresponding diffraction order m. Since the resonant angular positions are determined by the wavelength, the spectral information can be derived using the data plots. This measurement was performed only on a device surface, so the proposed method did not require an optical path to disperse the light.

3. Spectral Calculation Procedures

The spectral calculation basically followed the procedures presented in Reference [9]. We first presented the basic procedures and expanded them for a wider range of wavelengths. To calculate the spectrum, we first experimentally derived the light detection sensitivity of the photodetector called the responsivity R, which expresses the conversion efficiency from the input light intensity P_{in} into a measured photocurrent I_{ph}, defined as

$$R = \frac{I_{ph}}{P_{in}} \tag{2}$$

Since the photocurrent is measured during θ alteration, the responsivity R can be determined at each angle of incidence θ for monochromatic light irradiation as $R_{\lambda\theta} = I_\theta/P_\lambda$, where λ is the wavelength of the light. When light composed of several monochromatic lights of different wavelengths is simultaneously illuminated, the generated photocurrent is assumed to be a linear summation of the photocurrent responses of each wavelength component. Assuming that only three different wavelength lights, λ_1, λ_2, and λ_3, as in Figure 1c, are incident on the photodetector, the photocurrent measured at the SPR angle for λ_1, i.e., θ_1, becomes $I_{\theta_1} = R_{\lambda_1\theta_1}P_{\lambda_1} + R_{\lambda_2\theta_1}P_{\lambda_2} + R_{\lambda_3\theta_1}P_{\lambda_3}$. The measured value includes contributions from two other light wavelengths, λ_2 and λ_3. Under the same irradiation conditions, if the photocurrents were measured at two other SPR angles of incidence θ_2 and θ_3, the relationship between the photocurrent and incident intensity is expressed in a matrix form as,

$$\begin{bmatrix} I_{\theta_1} \\ I_{\theta_2} \\ I_{\theta_3} \end{bmatrix} = \begin{bmatrix} R_{\lambda_1\theta_1} & R_{\lambda_2\theta_1} & R_{\lambda_3\theta_1} \\ R_{\lambda_1\theta_2} & R_{\lambda_2\theta_2} & R_{\lambda_3\theta_2} \\ R_{\lambda_1\theta_3} & R_{\lambda_2\theta_3} & R_{\lambda_3\theta_3} \end{bmatrix} \begin{bmatrix} P_{\lambda_1} \\ P_{\lambda_2} \\ P_{\lambda_3} \end{bmatrix}. \tag{3}$$

This relationship can be extended to a larger matrix. Let λ_n be the wavelength range discretized into n components, I_n be the current of the SPR peak value, θ_n be the incident angle of the peak value, and the light intensity corresponding to each wavelength component λ_n be P_n. We obtain the following expression:

$$\begin{bmatrix} I_{\theta_1} \\ \vdots \\ I_{\theta_n} \end{bmatrix} = \begin{bmatrix} R_{\lambda_1\theta_1} & \cdots & R_{\lambda_n\theta_1} \\ \vdots & \ddots & \vdots \\ R_{\lambda_1\theta_n} & \cdots & R_{\lambda_n\theta_n} \end{bmatrix} \begin{bmatrix} P_{\lambda_1} \\ \vdots \\ P_{\lambda_n} \end{bmatrix}. \tag{4}$$

In short, $I = RP$, where P vector components indicate the intensity for each wavelength, and I vector components correspond to the measured photocurrents at each SPR angle of incidence. Since the diagonal components $R_{\lambda_i\theta_i}$ ($1 \leq i \leq n$) of the responsivity matrix R take a maximum among each column, the responsivity matrix R becomes regular with an inverse matrix. Because the P vector can be calculated by the equation $P = R^{-1}I$, the spectrum of the incident light can be derived.

When the operating wavelength range is narrow, the amplitude of the resonant angular position shift becomes narrow, and only a single SPR peak appears for each wavelength, as shown in a square of dotted lines in Figure 1c. If the wavelength range is expanded to cover the NIR wavelength range, the angular scanning range should also be expanded. Assuming that $\lambda_3 < \lambda_4 < \lambda_5$, multiple SPR peaks corresponding to different diffraction orders m may appear in the scanning range, as shown in the solid line square in Figure 1c. For example, with respect to the response to λ_1, not only θ_1 but also θ_9 peak appears. If we maintain a strategy to measure the photocurrents at SPR angles, the matrix takes the following form:

$$
\begin{bmatrix} I_{\theta_1} \\ \vdots \\ I_{\theta_9} \end{bmatrix} = \begin{bmatrix} R_{\lambda_1\theta_1} & \cdots & R_{\lambda_5\theta_1} \\ \vdots & \ddots & \vdots \\ R_{\lambda_1\theta_9} & \cdots & R_{\lambda_5\theta_9} \end{bmatrix} \begin{bmatrix} P_{\lambda_1} \\ \vdots \\ P_{\lambda_5} \end{bmatrix},
\tag{5}
$$

so that the R matrix is not a square matrix. Therefore, it is impossible to perform a calculation using a simple inverse matrix. Therefore, as a simple extension, we adopted a method of constructing an R matrix using the measured current values at all SPR peak angles occurring in the angular scanning range and reconstructed the incident spectrum using the generalization inverse matrix. When a generalized inverse matrix is used, the spectral derivation can be applied for a wider range of wavelengths. In summarized form, the spectral derivation equations become:

$$
\begin{bmatrix} I_{\theta_1} \\ \vdots \\ I_{\theta_k} \end{bmatrix} = \begin{bmatrix} R_{\lambda_1\theta_1} & \cdots & R_{\lambda_n\theta_1} \\ \vdots & \ddots & \vdots \\ R_{\lambda_1\theta_k} & \cdots & R_{\lambda_n\theta_k} \end{bmatrix} \begin{bmatrix} P_{\lambda_1} \\ \vdots \\ P_{\lambda_n} \end{bmatrix},
\tag{6}
$$

and

$$
P = \left(R^{\mathrm{T}}R\right)^{-1}R^{\mathrm{T}}I,
\tag{7}
$$

where R^{T} is a transpose of the R matrix. In the following, we experimentally constructed an R matrix and evaluated the applicability of the above method for NIR spectroscopy.

4. Materials and Methods

The design of the photodetector device was slightly modified from that in the previous report [9]. The substrate of the device was an n-Si wafer whose resistivity was 10–20 Ω·cm. One-dimensional gratings with 3.46-μm-pitch (denoted as a in Figure 1a) and height h of 40 nm were formed on the front surface of the n-Si substrate with an area size of 12.7 mm × 12.7 mm by reactive ion etching. The surface had a 100 nm thick Au film caused by vacuum evaporation during rotation with an oblique angle such that the sidewalls of the grating were covered in gold. On the back side of the n-Si substrate, an Al film was also formed as a cathode electrode. A photograph of the device is shown in Figure 2a. To confirm the diode rectification and photodetection characteristics, the current-voltage curve was measured (Figure 2b). The curve presented a typical diode characteristic. The Schottky barrier height was calculated following a procedure described in Reference [16], and was 0.776 eV. Since the detection limit wavelength of the Schottky barrier height corresponds to 1.62 μm, near-infrared light detection was confirmed to be possible. In addition, a surface topographic image was taken using an AFM (JSPM-5000, JEOL, Tokyo, Japan). It was confirmed that a grating with a height amplitude of approximately 40 nm was fabricated. Since obtainable photocurrent signals for spectral measurements were on the order of nA to μA, the signal often suffered from noise. The current was therefore converted and amplified to a voltage in the immediate vicinity of the device using an operational amplifier to improve signal-to-noise ratio and signal resolution (Figure 2c). The first operational amplifier converted the output current I_{ph} to a voltage through a feedback resistor $R = 1\ \mathrm{M\Omega}$ according to an equation below:

$$
V_o = I_{\mathrm{ph}} \times R.
\tag{8}
$$

Then, the converted voltage V_o was passed to a voltage follower to provide the output voltage V_{out}. In the following, the current values were calculated by dividing V_{out} by $R = 1$ MΩ. The conversion amplifier and the photodetector device were put into a shielding box, and an aperture was made in a box wall in front of the photodetector such that the light was incident on the photodetector, as shown in Figure 2a.

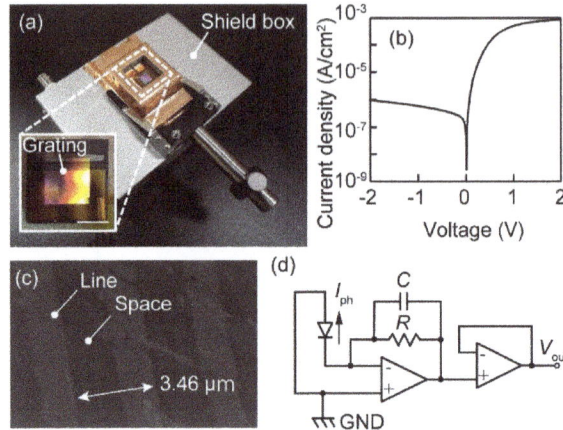

Figure 2. Photographs of the fabricated photodetector and the measurement circuit unit (bar = 5 mm). (**a**) The fabricated photodetector in a shield box. (**b**) A characteristic current-voltage curve of the photodetector. (**c**) An AFM image of the grating area. (**d**) An I-V conversion amplifier circuit.

5. Experimental Results

A responsivity matrix R was experimentally constructed. The experimental setup is shown in Figure 3a. In the measurement, the device was fixed on a rotational stage, and infrared light from a wavelength tunable laser (SC-450, Fianium, Southampton, UK) was incident on the device. The wavelengths of the monochromatic infrared light were scanned from 1200 nm to 1600 nm with a 25-nm interval of wavelength. A linear polarizer was installed between the light source and the device such that the light became TM-polarized. The photocurrent was converted to a voltage signal by an I-V conversion amplifier and was measured using a source meter (6242, ADCMT, Tokyo, Japan). Using the rotating stage, the angle of incidence θ was rotated from 0° to 30° with a resolution of 0.1°. The light intensity at each wavelength was measured using a power meter (PM300, Thorlabs, Newton, NJ, USA). Experiments were conducted in a darkroom to prevent stray light. The obtained photocurrent angular spectra are shown in Figure 3b, where the SPR peaks are highlighted with arrows. The vertical axis is the logarithmic representation of the responsivity, and the horizontal axis is the angle of incidence. Since the laser spot area was smaller than the grating area, all of the incident energy was used to calculate the responsivity. Noise reduction due to the I-V amplifier provided clear photocurrent waveforms in a wide wavelength range. The SPR peak angular positions systematically shifted depending on the wavelength. Since the angular scanning range was wider than that in a previous report [9], multiple SPR peaks corresponding to different diffraction orders m were found in each curve. It was also found that a photocurrent was generated even when SPR did not occur. This baseline photocurrent can be attributed to excitation of electrons by the direct irradiation of the near-infrared light on the Au/n-Si Schottky interface through the Au film. This photocurrent presented a tendency to decrease with an increase in the wavelength because the photon energy of the light was inversely proportional to the wavelength. Elimination of this baseline photocurrent by optimization of device structures will be necessary to improve the spectrometer performance in the future.

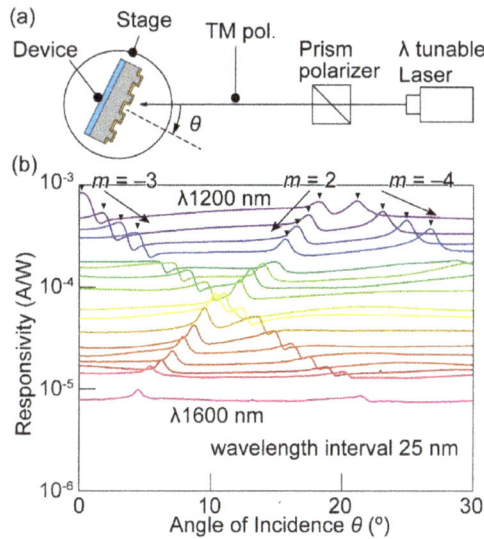

Figure 3. (**a**) An experimental setup. (**b**) Responsivity angular spectrum for near-infrared (NIR) monochromatic laser irradiation for wavelengths ranging from 1200 nm to 1600 nm.

To investigate the validity of the obtained angular spectrum, the SPR peak angular positions were compared with the calculated ones using Equation (1). The dispersive permittivity of the gold was taken from Reference [17] in the calculation. The calculated theoretical SPR angles were plotted with respect to the wavelength in addition to the measured angles in Figure 4. The calculated and experimental SPR angular positions showed high consistency. Although there was some error in the angle, particularly for $m = -3$, the amplitude was as small as 1°. It was therefore concluded that the obtained photocurrent peaks can be attributed to SPR. The responsivity matrix R was then constructed using responsivity values at SPR peak angular points belonging to all three diffraction orders.

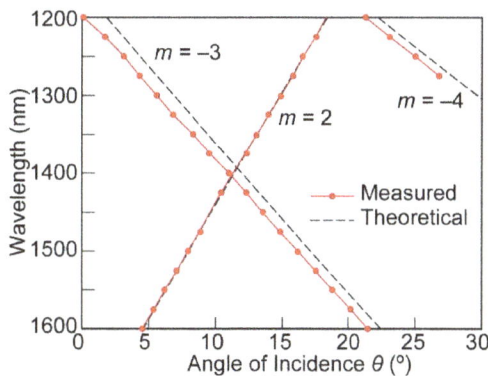

Figure 4. SPR angle positions with respect to the wavelength of the incident light. Measured and calculated theoretical angle positions belonging to three different diffraction orders.

To check the spectroscopy performance using an R matrix with a generalized inverse matrix method, light composed of two wavelengths, 1250 nm and 1450 nm, was irradiated at the same time, and the spectrum was calculated. The results are shown in Figure 5a.

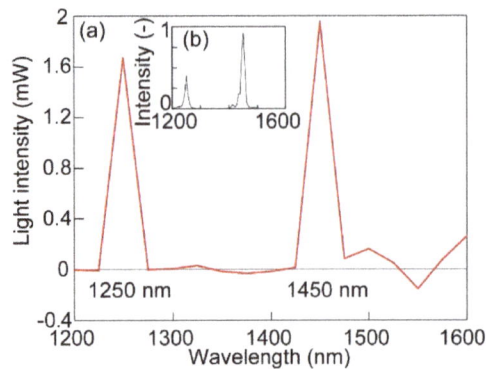

Figure 5. Spectrum of the light composed of two different wavelengths, 1250 nm and 1450 nm. (**a**) Calculated spectral data obtained with the fabricated device in this article, (**b**) spectrum of the light obtained with a commercial spectrometer.

The horizontal and vertical axes indicate the wavelength and intensity, respectively. Distinctive peaks were observed at $\lambda = 1250$ nm and 1450 nm. The intensity at these two peak wavelengths was consistent with the values measured with a power meter. The reference spectrum data in Figure 5b were measured with a commercially available NIR compact spectrometer (Sol. 2.2A, B&W Tek, Newark, DE, USA). Peak positions and spectrum shapes were consistent between the two. This spectral data consistency can be attributed to the fact that the inverse matrix method takes contributions from all diffraction orders into consideration. Therefore, NIR spectroscopy was performed using the responsivity data ranging from $\lambda = 1200$ nm to 1600 nm. There is, however, a difference on the power peak height between Figure 5a,b. It may be attributed to the laser intensity fluctuation because these two data were obtained at different times. Quantitative evaluation of the proposed sensor performance should further be performed in the future.

6. Conclusions

In this paper, we proposed NIR spectroscopy based on a Si photodetector equipped with a gold grating and extended the measurable NIR wavelength range by improving a spectrum derivation procedure from a previous study [9]. A responsivity matrix R was constructed, and a generalized inverse matrix of R was used to obtain the spectrum to fit a situation where multiple SPR peaks appeared among the scanning range. When light composed of two wavelengths, 1250 nm and 1450 nm, was irradiated at the same time, the two wavelengths were distinctively discriminated using the improved method. The reduction of the angular scanning resolution will provide a denser R matrix to improve the wavelength resolution. Moreover, since sharpening of an SPR peak curve shape can be realized by tuning the grating profile, the wavelength resolution can further be improved [18–20]. It is noted that the improvement of sensitivity should be pursued for the use with a normal white light source instead of a laser as a light source because the responsivity of the proposed device is still around 10–100 µA/W. Since the previous literature indicates that responsivity as large as several mA/W is possible with the similar plasmonic and Schottky approach [21], two-orders of improvement can be expected, which will provide us with spectrum data with practically high signal-to-noise ratio. The proposed spectrometer device is made of Si, so it is possible to integrate the angular scanning mechanism into the photodetector as a micro-electro-mechanical systems (MEMS) device [22,23]. The proposed spectroscopy method enables spectroscopy with only a thin-film plasmonic layer and an integrated photodetector located beneath the layer, and it is advantageous from the viewpoint of miniaturization. Advancement of the proposed method will provide a new microsized integrated spectrometer that will provide rich information about our environment.

Micromachines **2019**, *10*, 403

Author Contributions: T.K. designed the research, Y.S. performed the research, Y.Y. measured the responsivity data, G.T. implemented the measurement software, Y.A. analyzed the Schottky data, T.K. and Y.S. wrote the paper.

Funding: This research was partially supported by the Tateishi Science and Technology Foundation, Japan.

Acknowledgments: The EB direct writings were carried out using the EB lithography apparatus of the VLSI Design and Education Center (VDEC) of the University of Tokyo. The microfabrications were performed in a clean room of the Division of Advanced Research Facilities (DARF) of the Coordinated Center for UEC Research Facilities of the University of Electro-Communications, Tokyo, Japan.

Conflicts of Interest: The authors declare no conflict of interest.

References

1. Magwaza, L.S.; Opara, U.L.; Nieuwoudt, H.; Cronje, P.J.R.; Saeys, W.; Nicolaï, B. NIR Spectroscopy applications for internal and external quality analysis of citrus fruit—A review. *Food Bioprocess. Technol.* **2012**, *5*, 425–444. [CrossRef]
2. Nicolaï, B.M.; Beullens, K.; Bobelyn, E.; Peirs, A.; Saeys, W.; Theron, K.I.; Lammertyn, J. Nondestructive measurement of fruit and vegetable quality by means of NIR spectroscopy: A review. *Postharvest Biol. Technol.* **2007**, *46*, 99–118. [CrossRef]
3. Roggo, Y.; Chalus, P.; Maurer, L.; Lema-Martinez, C.; Edmond, A.; Jent, N. A review of near infrared spectroscopy and chemometrics in pharmaceutical technologies. *J. Pharm. Biomed. Anal.* **2007**, *44*, 683–700. [CrossRef]
4. Villringer, A.; Chance, B. Non-invasive optical spectroscopy and imaging of human brain function. *Trends Neurosci.* **1997**, *20*, 435–442. [CrossRef]
5. Tsur, Y.; Arie, A. On-chip plasmonic spectrometer. *Opt. Lett.* **2016**, *41*, 3523. [CrossRef] [PubMed]
6. Fleischman, D.; Sweatlock, L.A.; Murakami, H.; Atwater, H. Hyper-selective plasmonic color filters. *Opt. Express* **2017**, *25*, 27386. [CrossRef] [PubMed]
7. Ema, D.; Kanamori, Y.; Sai, H.; Hane, K. Plasmonic color filters integrated on a photodiode array. *Electron. Commun. Jpn.* **2018**, *101*, 95–104. [CrossRef]
8. Kim, S.; Lee, Y.; Kim, J.Y.; Yang, J.H.; Kwon, H.J.; Hwang, J.Y.; Moon, C.; Jang, J.E. Color-sensitive and spectrometer-free plasmonic sensor for biosensing applications. *Biosens. Bioelectron.* **2019**, *126*, 743–750. [CrossRef]
9. Chen, W.; Kan, T.; Ajiki, Y.; Matsumoto, K.; Shimoyama, I. NIR spectrometer using a Schottky photodetector enhanced by grating-based SPR. *Opt. Express* **2016**, *24*, 25797. [CrossRef]
10. Raether, H. *Surface Plasmons on Smooth and Rough Surfaces and on Gratings*; Springer: Berlin, Germany, 1988; Volume 111.
11. Kimata, M.; Denda, M.; Yutanj, N.; Iwade, S.; Tsubouchi, N. A 512 × 512-element PtSi Schottky-barrier infrared image sensor. *IEEE J. Solid-State Circuits* **1987**, *22*, 1124–1129. [CrossRef]
12. Kosonocky, W.F.; Shallcross, F.V.; Villani, T.S.; Groppe, J.V. 160 × 244 element PtSi Schottky-barrier IR-CCD image sensor. *IEEE Trans. Electron. Devices* **1985**, *32*, 1564–1573. [CrossRef]
13. Scales, C.; Berini, P. Thin-film Schottky barrier photodetector models. *IEEE J. Quantum Electron.* **2010**, *46*, 633–643. [CrossRef]
14. Casalino, M.; Sirleto, L.; Moretti, L.; Gioffrè, M.; Coppola, G.; Rendina, I. Silicon resonant cavity enhanced photodetector based on the internal photoemission effect at 1.55 μm: Fabrication and characterization. *Appl. Phys. Lett.* **2008**, *92*, 251104. [CrossRef]
15. Sze, S.M.; Ng, K.K. *Physics of Semiconductor Devices*; John Wiley & Sons: Hoboken, NJ, USA, 2006.
16. Cheung, S.K.; Cheung, N.W. Extraction of Schottky diode parameters from forward current-voltage characteristics. *Appl. Phys. Lett.* **1986**, *49*, 85–87. [CrossRef]
17. Rakić, A.D.; Djurišić, A.B.; Elazar, J.M.; Majewski, M.L. Optical properties of metallic films for vertical-cavity optoelectronic devices. *Appl. Opt.* **1998**, *37*, 5271–5283. [CrossRef] [PubMed]
18. Cai, D.; Lu, Y.; Lin, K.; Wang, P.; Ming, H. Improving the sensitivity of SPR sensors based on gratings by double-dips method (DDM). *Opt. Express* **2008**, *16*, 14597. [CrossRef]
19. Yoon, K.H.; Shuler, M.L.; Kim, S.J.; Jung, J.-M.; Shin, Y.-B.; Kim, M.-G.; Ro, H.-S.; Jung, H.-T.; Chung, B.H. Design optimization of nano-grating surface plasmon resonance sensors. *J. Phys. Chem. B* **2006**, *14*, 22351–22358. [CrossRef]

20. Pi, S.; Zeng, X.; Zhang, N.; Ji, D.; Chen, B.; Song, H.; Cheney, A.; Xu, Y.; Jiang, S.; Sun, D.; et al. Dielectric-grating-coupled surface plasmon resonance from the back side of the metal film for ultrasensitive sensing. *IEEE Photonics J.* **2016**, *8*, 4800207.

21. Li, W.; Valentine, J. Metamaterial perfect absorber based hot electron photodetection. *Nano Lett.* **2014**, *14*, 3510–3514. [CrossRef] [PubMed]

22. Ford, J.E.; Aksyuk, V.A.; Bishop, D.J.; Walker, J.A. Wavelength add-drop switching using tilting micromirrors. *J. Light. Technol.* **1999**, *17*, 904–911. [CrossRef]

23. Dickensheets, D.L.; Kino, G.S. Silicon-micromachined scanning confocal optical microscope. *J. Microelectromech. Syst.* **1998**, *7*, 38–47. [CrossRef]

micromachines

MDPI

Article

Magneto-Impedance Sensor Driven by 400 MHz Logarithmic Amplifier

Tomoo Nakai

Industrial Technology Institute, Miyagi Prefectural Government, Sendai 981-3206, Japan;
nakai-to693@pref.miyagi.lg.jp; Tel.: +81-22-377-8700

Received: 30 April 2019; Accepted: 28 May 2019; Published: 29 May 2019

Abstract: A thin-film magnetic field sensor is useful for detecting foreign matters and nanoparticles included in industrial and medical products. It can detect a small piece of tool steel chipping or breakage inside the products nondestructively. An inspection of all items in the manufacturing process is desirable for the smart manufacturing system. This report provides an impressive candidate for realizing this target. A thin-film magneto-impedance sensor has an extremely high sensitivity, especially, it is driven by alternatiing current (AC) around 500 MHz. For driving the sensor in such high frequency, a special circuit is needed for detecting an impedance variation of the sensor. In this paper, a logarithmic amplifier for detecting a signal level of 400 MHz output of the sensor is proposed. The logarithmic amplifier is almost 5 mm × 5 mm size small IC-chip which is widely used in wireless devices such as cell phones for detecting high-frequency signal level. The merit of the amplifier is that it can translate hundreds of MHz signal to a direct current (DC) voltage signal which is proportional to the radio frequency (RF)signal by only one IC-chip, so that the combination of a chip Voltage Controlled Oscillator (VCO), a magneto-impedance (MI) sensor and the logarithmic amplifier can compose a simple sensor driving circuit.

Keywords: magneto-impedance sensor; thin-film; high frequency; logarithmic amplifier; nondestructive inspection

1. Introduction

A thin-film magneto-impedance sensor [1–6] is useful for detecting magnetic materials nondestructively. The sensor has high sensitivity and also has tolerance of normal magnetic field because of its demagnetizing force in the thickness direction. Our previous report showed that this sensor has sensitivity of 1.7×10^{-9} T at 300 Hz inside the normal field of 0.1 T [7]. By using this sensor, a single magnetic particle with 65 μm diameter could be detected in the distance 0.5 mm above the sensor plane. In this measurement, the output signal was over 6 V out of the measurement of the single 65 μm particle with a low remanence property. The point is that it was detected with subjecting a static normal field of 0.1 T around a measurement area including the thin-film sensor. Based on the previous work, we are trying to extend the application of this sensor system. The aim of this study is developing a method of nondestructive detection of magnetic small bodies such as foreign matters included in industrial products and also detection of a density of magnetic nanoparticles in a drug solution. In the case of the industrial application, it can detect a small piece of tool steel chipping or breakage inside the products nondestructively. An inspection of all items in the manufacturing process is desirable for the smart manufacturing system. In order to apply an actual manufacturing process a system without magnetic shielding is desirable due to its low cost and small installation space of equipment. An environmental noise in a factory which overwhelms the detected signal can be solved by making the detection signal stronger using a stronger normal magnetic field. The thinner the sensor thickness the more tolerance appears against strong normal field, because of the demagnetizing field. For driving such thin sensor, high frequency is essential for the skin-effect of magneto-impedance. For the measurement system which applied for a conveyer in manufacturing process and medical application the signal frequency range is expected to be from DC to several Hz due to the movement speed of products or scanning speed of measurement probe. This paper provides an impressive candidate for realizing this target. A $Co_{85}Nb_{12}Zr_3$ thin-film magneto-impedance sensor has an extremely high sensitivity, especially, it is driven by AC current around 500 MHz [8]. For driving the sensor in such high frequency, a special circuit is needed for detecting an impedance variation of the sensor. Previous works which report high frequency magneto-impedance [9–11] used impedance-analyzer, network-analuzer and signal-analyzer for detection. In this paper, an application of a logarithmic amplifier for detecting a level of the sensor signal at 400 MHz, which is proportional to the sensor impedance, is proposed. The logarithmic amplifier is almost 5 mm × 5 mm size small IC-chip which is widely used in wireless devices such as cell phones for detecting high-frequency signal level. The merit of the amplifier is that it can transfer hundreds of MHz signal to a DC voltage signal which is proportional to the RF signal by only one IC-chip, so that the combination of a chip Voltage Controlled Oscillator (VCO), a MI sensor and the logarithmic amplifier can compose a simple sensor driving circuit. By using the logarithmic amplifier, the output of the sensor circuit suited for a frequency range from DC to several Hz. The sensing circuit has a configuration of differential input for the purpose of getting high sensitivity. One of the inputs is a sensor signal and the other is a reference signal which has the same phase and the amplitude as the one where an external magnetic field is a certain reference value. The output frequency bandwidth of the logarithmic amplifier was ranging from DC to 20 kHz. But in this study the output frequency range was limited in which a signal of measured object carried on a conveyer. This frequency was under several Hz, and the system will be designed without using a magnetic field shield structure. The most sensitive frequency range of our sensor element was around 500 MHz, due to the element sensitivity dZ/dH marks a maximum around 500 MHz, here Z is element impedance and H is external field in the sensing direction. Whereas the circuit in this study was 400 MHz. The reason is a selection of devices which are commercially accessible. If a suitable device would be getting accessible the 500 MHz driving circuit can be made using the same design rule as this paper.

2. Experimental Procedures and Results

2.1. Sensor Element and Driving Circuit

Figure 1 shows the view of sensor element. The element was fabricated by a thin-film process. An amorphous $Co_{85}Nb_{12}Zr_3$ film was RF-sputter deposited onto a soda glass substrate and then micro-fabricated into rectangular elements by a lift-off process. The element was 1000 µm length, 50 µm wide, 2.1 µm thick. The tens of elements are aligned in a parallel configuration and connected by Cu thin-film strips to form a meander pattern. A magnetic field was applied while the RF-sputter deposition for the purpose of inducing uniaxial magnetic anisotropy. The direction of the magnetic anisotropy in this study was in the width direction, Y, so to say short side direction of the element strip. It is induced by the direction of the magnetic field while sputtering. The sensor element was mounted on printed circuit board (PCB) and electrically connected by a conductive silver paste.

A typical example of variation of impedance of the sensor itself without a PCB as a function of external field is shown in Figure 2. In this figure, the external field was applied in the in-plane length direction of the sensor strip, X-direction, which is the sensing direction. This impedance variation was obtained by S11 measurement of a network-analyzer using high-frequency probe for electric connection with the electrode pads of sensor element. The result shows a parabolic variation, which is typical for a magneto-impedance sensor having magnetic anisotropy in the width direction. The minimum impedance, |Z|=138 Ω at 0 kA/m (0 Oe), and the maximum was 275 Ω around ±1.76 kA/m (±22 Oe). In this figure, a bias point is shown, on which this sensor was operated.

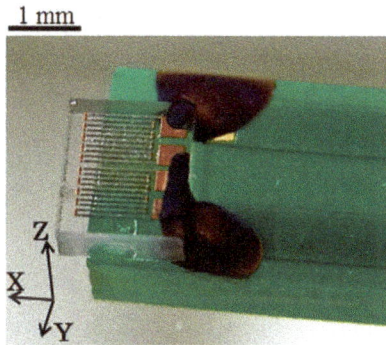

Figure 1. View of the sensor element.

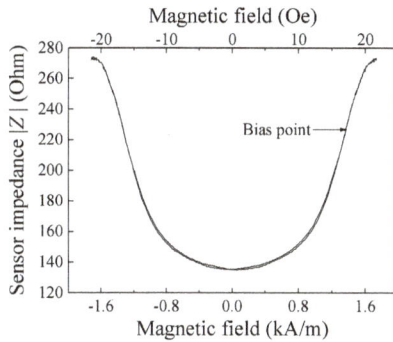

Figure 2. Typical example of variation of impedance of the sensor.

Figure 3 shows the magnetic domain of the sensor element observed by Kerr microscope (BH-762PI-MAE, NEOARK Corporation, Tokyo, Japan). The magnetic domain forms contiguous regions of magnetic momentum with anti-parallel direction. This picture shows a magnetic domain when an external magnetic field was zero. As increasing or decreasing the external field applied along the length direction, X, which is the sensing direction of sensor, the width of the contiguous domain regions changes and finally the domain would be a single domain with a momentum directing the same as the external field direction. In this element, the magnetic momentum in each region vibrates by the effect of the high-frequency current running through the element, and the impedance of the element changes as the magnetic domain changes. In another word, the magneto-impedance effect comes from the change of the condition of the momentum vibration caused by the change of both the direction and the distribution of the momentum in the thin-film element having an in-plane uniaxial magnetic anisotropy.

Figure 4 shows a schematic of magnetic field when the sensor is operated. The sensor needs a biasing field for operation in the X direction, it is about 1.4 kA/m (17.5 Oe) as shown in Figure 2. The sensing magnetic field is also in X direction. Therefore, a constant bias field is essential for this sensor system.

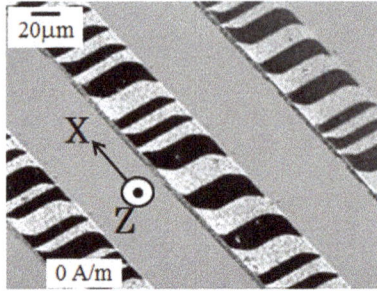

Figure 3. Magnetic domain of the sensor element observed by Kerr microscope.

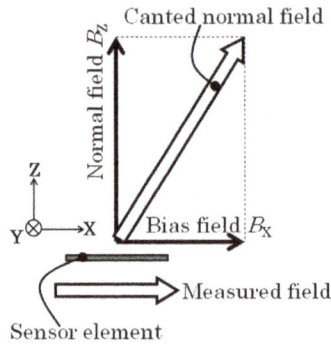

Figure 4. Schematic of magnetic field when the sensor is operated.

Figure 5 shows the proposed driving circuit of the thin-film sensor. A 400 MHz alternating signal at −5.3 dBm was generated by a voltage-controlled oscillator. The signal was divided into two and one was introduced to the sensor element and the other was introduced to a series of RF control devices. As shown in the figure the latter is a series of variable attenuator and a variable phase shifter. Both of them were electrically voltage-controlled ones. The RF signal going through the latter way of circuit branch was controlled in the same phase and amplitude as the one going through the sensor when an external magnetic field is a certain reference value. Each signal was inputted to a logarithmic amplifier as a pair of differential inputs. The logarithmic amplifier detects 400 MHz signal into a voltage signal logarithmically proportional to the signal level. A final output of the circuit was processed with an offset compensation and an amplifying 100 times. The baseline stability of the output is satisfactory even its high-sensitivity.

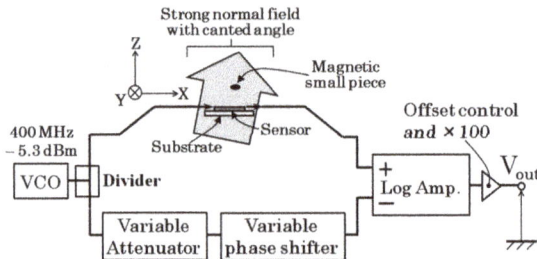

Figure 5. Proposed driving circuit of the thin-film sensor.

2.2. Results of Experiment

At first, a certainty of the output of this sensor system is confirmed. The system was composed of the sensor on PCB (Figure 1) and driving circuit (Figure 6). It is confirmed by measuring the output signal of the system caused by the change of the element impedance. This measurement was carried out by applying magnetic field in the X direction to the sensor on PCB with driving it by the circuit. The variation of element impedance without PCB is previously shown in Figure 2. The driving circuit was the same as Figure 5 but the attenuator was set to be maximum attenuation. This setting makes an input of the '−' terminal of the logarithmic amplifier to be almost zero, then the output was expected to be logarithmically proportional to the sensor impedance.

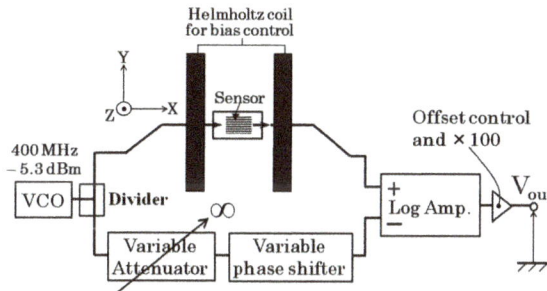

Figure 6. Method of confirmation of certainty of the output of sensor system.

Figure 7 shows a result of measurement. The external magnetic field was applied in X direction.

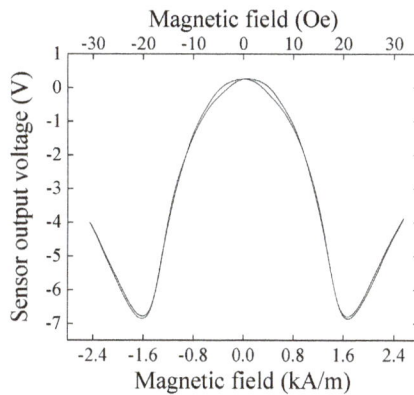

Figure 7. Result of sensor impedance measurement by the circuit.

In this measurement, the normal magnetic field was in a room condition of Japanese north-east region. The output voltage was ranging from −7 V to +0.2 V and it was in good agreement with the impedance profile of the sensor element. In our measurement the baseline level was an arbitrary one because a level of offset compensation in the final circuit stage was arbitrary. The profile of sensor output (Figure 7) was a vertical inversion of the element impedance (Figure 2). The reason is that the RF level come out from sensor element is inversely proportional to its impedance.

Figure 8 shows measured results when a normal magnetic field, B_z = 1,100 G (0.11 T), was applied to the sensor element. In this measurement, the external magnetic field was applied in X direction. As the normal field increases, the vertical range of variation slightly decreases even it is keeping a low hysteresis of impedance variation. It means the sensitivity dV/dH does not decrease even in 0.11

T normal field, where V is the output voltage and H is the magnetic field. Here the normal field was applied by NdFeB magnets placed both on the upper side and on the lower side of the sensor element. These magnets were attached on an opening tip of a C-shaped magnetic core made by a bundle of silicone-steel sheets. The external field in the sensing direction was applied by a Helmholtz coil. A photograph of the measurement apparatus is shown in Figure 9. This result shows that the thin-film magneto-impedance sensor has a possibility to be useable even in hundreds of mT normal magnetic field and a strong candidate for detecting foreign inclusion in the manufacturing process.

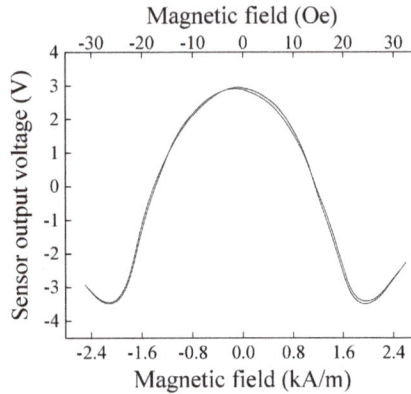

Figure 8. Measured result when a normal magnetic field $B_Z = 0.11$ T.

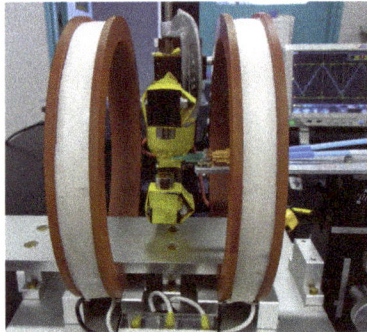

Figure 9. Photograph of the measurement apparatus with applying normal field.

Now it proceeds to an evaluation of output property and sensitivity of the sensor system. This confirmation of sensing property was carried out without normal field. Figure 10 shows a schematic of measurement apparatus which was used in this experiment. Figure 11 shows a photograph of dual-Helmholtz coil equipment used in this study. One of the Helmholtz coil was used for applying a DC bias magnetic field and the other was used for generating a small AC magnetic field for the purpose of sensing limit evaluation. The biasing DC field was 1.4 kA/m (17.5 Oe) and the AC field was ranging from 0.5×10^{-7} T to 160×10^{-7} T. From here the applied magnetic field is assumed to be a value in vacuum, therefore it is expressed in a magnetic flux density in vacuum.

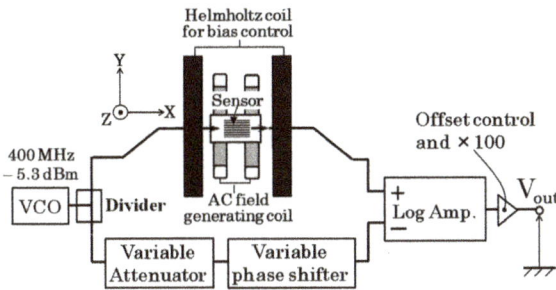

Figure 10. Measurement apparatus used in the sensitivity evaluation.

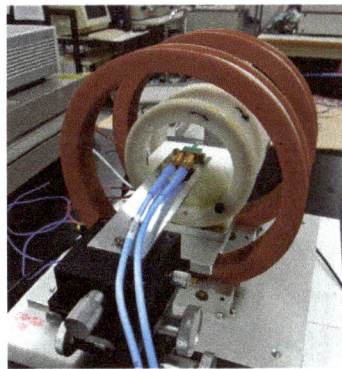

Figure 11. Photograph of dual-Helmholtz coil equipment used in sensitivity evaluation.

Figure 12 shows a variation of the sensor output as a function of AC magnetic flux density at 5 Hz applied to the sensor element. This log-log plot shows that it has approximately linear relation. Due to a limit of output level of the final-stage amplifier, up to +14 V, the maximum value of output was +14 V in this study. The system developing in this study is designed for use in manufacturing process without magnetic shielding. Due to it a circumferential magnetic noise would be approximately several mG. With consideration of a background noise of the driving circuit, the minimum signal level must be larger than 0.1 mV in amplitude level. The output level of the developed driving circuit conforms to this design criterion. The AC field ranging from 1×10^{-7} T to 160×10^{-7} T corresponds to the output level ranging from 0.1 V to 14 V.

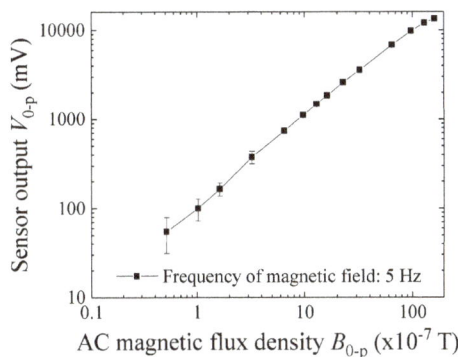

Figure 12. Variation of the sensor output as a function of AC magnetic flux density at 5 Hz.

The sensor sensitivity was measured by using spectrum analyzer. The sensitivity was defined as a minimum limit of the amplitude of magnetic field where a peak of the signal sunk under a noise level of the spectrum measurement. In this measurement, a DC-cut filter and a 20 dB attenuator was used for the reason of minimizing the noise level around DC and protect from over power. A Real Time Spectrum Analyzer (RTSA) was used for this measurement due to its ability of low frequency spectrum measurement including DC. The Tektronix RSA3408A was used for this measurement.

Figure 13 shows a measurement result. The horizontal-axis shows a frequency from DC to 50 Hz, the vertical-axis shows a signal level of output which is including alternating magnetic field, 3.2×10^{-8} $T_{0\text{-}P}$ (0.32 m$G_{0\text{-}P}$) at 3 Hz. From this figure, the 3 Hz peak clearly sticks out above the noise level. In other words, the sensitivity of the sensor system achieves 3.2×10^{-8} T (0.32 mG).

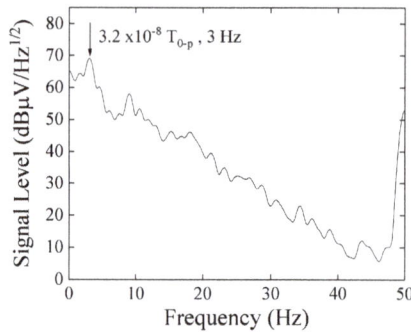

Figure 13. Result of spectrum measurement for a signal of 3.2×10^{-8} $T_{0\text{-}P}$ at 3 Hz.

3. Discussion

A discussion on the experimentally obtained sensitivity using the proposed driving circuit and a future subject of this work would be carried out here.

In this section, a cause of background noise and a relevance of achieved sensitivity are discussed.

Figure 14 shows a noise level obtained for measurement apparatus RSA3408A only. The input was terminated by 50 Ω and the sensor system was not connected. Figure 15 shows a noise level with connection of the sensor unit without AC magnetic field. With comparison of them, the noise of measurement apparatus (RSA3408A, Tektronix) was lower than the one with sensor unit. The 50 Hz noise in Figures 13 and 15 can be estimated as the Japanese commercial power supply signal. The 1/f-noise observed in the range from DC to 50 Hz would be estimated as a combination of system noise of the sensor driving circuit and circumferential magnetic field in the laboratory. The ratio of strength of these noises could be obtained by a measurement using magnetic shielding, but in my lab. we have not it.

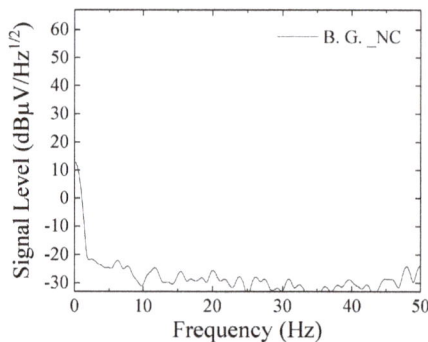

Figure 14. Noise level of measurement apparatus RSA3408A only.

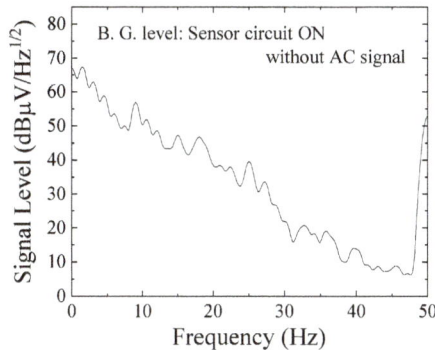

Figure 15. Noise level with connection of the sensor unit without AC magnetic field.

Figure 16 shows an actual time-domain measurement result of the noise from the sensor system. This result was obtained using the sensor driving circuit without AC magnetic field. The 0.67 s interval dip-point was observed as a periodically repeated waveform with constant baseline, and it seems that it comes from the driving circuit. The small sine-signal is come from the commercial power supply because of the frequency 50 Hz. The previous report [7] was measured without using the electrically controlled attenuator and phase-shifter. It also used narrow band-pass filter at 300 Hz. If the cause of noise in the circuit will be cleared, the sensitivity will have a possibility to go up nearly 1.7×10^{-9} T.

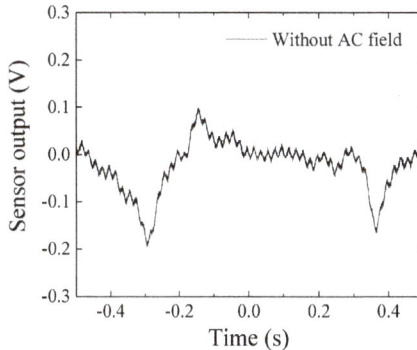

Figure 16. Measured time-domain noise of the sensor system.

Figure 17 shows a measured signal when a 5.1×10^{-8} T_{0-P} in 5 Hz was applied to the sensor element. We can see the periodical sine signal in this measurement. It is natural that this sine-signal is combined with the system-noise (Figure 16), but the 0.2 s periodical peak is clearly observed.

The achieved sensitivity 3.2×10^{-8} T_{0-P} shown in the results may be reasonable based on Figure 17. We can easily understand that a reduction of the system noise and 50 Hz noise could make the sensitivity drastically improve. An estimation of system sensitivity with application of the strong normal magnetic field is a future subject. The sensor-head apparatus shown in Figure 9 suffered a vibration coming from both the C-shaped core and the sensor fixing metal-arm. The more rigid apparatus and larger electro-magnetic core for generating a normal field will be needed for the precise experiment. I am now preparing such apparatus and will soon make a report about it.

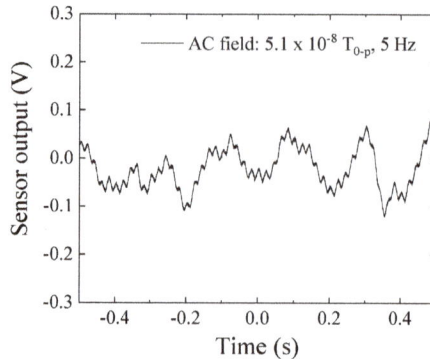

Figure 17. Measured signal when a 5.1×10^{-8} T_{0-P} in 5 Hz was applied to the sensor element.

4. Conclusion

A driving circuit for a thin-film magneto-impedance sensor using 400 MHz logarithmic amplifier is proposed. The highest sensitivity of the sensor used in this study is around 500 MHz, and the result of this article is well approaching and applicable to this frequency. An expected application of this sensor system is a nondestructive inspection of foreign matters in industrial products and detection of magnetic nanoparticles in drug solutions. For applying these targets, the frequency range of detection signal must be under several Hz. The proposed circuit using a logarithmic amplifier detects 400 MHz signal and make an output of the suitable frequency range. The achieved sensitivity of this system was 3.2×10^{-8} T at 3 Hz. This is a result without strong normal magnetic field, but it is expected to have nearly the same sensitivity in case of applying normal field because of the slight degradation of MI property of the sensor element (Figure 7). The sensitivity evaluation in a strong magnetic field will soon be reported in a future article.

Funding: This research received no external funding.

Conflicts of Interest: The authors declare no conflict of interest.

References

1. Mohri, K.; Bushida, K.; Noda, M.; Yoshida, H.; Panina, L.V.; Uchiyama, T. Magneto-Impedance Element. *IEEE Trans. Magn.* **1995**, *31*, 2455–2460. [CrossRef]
2. Uchiyama, T.; Mohri, K.; Panina, L.V.; Furuno, K. Magneto-impedance in sputtered amorphous films for micro magnetic sensor. *IEEE Trans. Magn.* **1995**, *31*, 3182–3184. [CrossRef]
3. Mohri, K.; Uchiyama, T.; Shen, L.P.; Cai, C.M.; Honkura, Y.; Aoyama, H. Amorphous wire and CMOS IC based sensitive micro-magnetic sensors utilizing magneto-impedance (MI) and stress-impedance (SI) effects and applications. In *MHS2001. Proceedings of 2001 International Symposium on Micromechatronics and Human Science (Cat. No. 01TH8583)*; IEEE: Piscataway, NJ, USA, 2001; pp. 25–34.
4. Raposo, V.; Vazquez, M.; Flores, A.G.; Zazo, M.; Iniguez, J.I. Giant magnetoimpedance effect enhancement by circuit matching. *Sens. Actuators A* **2003**, *106*, 329–332. [CrossRef]
5. Yabukami, S.; Suzuki, T.; Ajiro, N.; Kikuchi, H.; Yamaguchi, M.; Arai, K.I. A high frequency carrier-type magnetic field sensor using carrier suppressing circuit. *IEEE Trans. Magn.* **2001**, *37*, 2019–2021. [CrossRef]
6. Zhao, W.; Bu, X.; Yu, G.; Xiang, C. Feedback-type giant magneto-impedance sensor based on longitudinal excitation. *J. Magn. Magn. Mater.* **2012**, *324*, 3073–3077. [CrossRef]
7. Nakai, T. Study on detection of small particle using high-frequency carrier-type thin film magnetic field sensor with subjecting to strong normal field. The papers of Technical Meeting on Physical Sensor, IEE Japan, PHS-16-15. 2016, pp. 11–20. (In Japanese). Available online: http://id.nii.ac.jp/1031/00093680/ (accessed on 30 April 2019).

8. Nakai, T. Study on low bias GMI sensor with controlled inclined angle of stripe magnetic domain. Ph.D. Thesis, Tohoku University, Sendai, Japan, 2005.

9. Fernandez, E.; Garcia-Arribas, A.; Barandiarán, J.M.; Svalov, A.V.; Kurlyandskaya, G.V.; Dolabdjian, C.P. Equivalent Magnetic Noise of Micro-Patterned Multilayer Thin Films Based GMI Microsensor. *IEEE Sens. J.* **2015**, *15*, 6707–6714. [CrossRef]

10. Thiabgoh, O.; Shen, H.; Eggers, T.; Galati, A.; Jiang, S.; Liu, J.S.; Li, Z.; Sun, J.S.; Srikanth, H.; Phan, M.H. Enhanced high-frequency magneto-impedance response of melt-extracted $Co_{69.25}Fe_{4.25}Si_{13}B_{13.5}$ microwires subject to Joule annealing. *J. Sci. Adv. Mater. Devices* **2016**, *1*, 69–74. [CrossRef]

11. Chlenova, A.A.; Moiseev, A.A.; Derevyanko, M.S.; Semirov, A.V.; Lepalovsky, V.N.; Kurlyandskaya, G.V. Permalloy-Based Thin Film Structures: Magnetic Properties and Giant Magnetoimpedance Effect in the Temperature Range Important for Biomedical Applications. *Sensors* **2017**, *17*, 1900. [CrossRef] [PubMed]

micromachines   **MDPI**

Article

A Thermal Tactile Sensation Display with Controllable Thermal Conductivity

Seiya Hirai and Norihisa Miki *

Department of Mechanical Engineering, Keio University, 3-14-1 Hiyoshi, Kohoku-ku, Yokohama, Kanagawa 223-8522, Japan; seiya_hirai@keio.jp
* Correspondence: miki@mech.keio.ac.jp; Tel.: +81-45-566-1430

Received: 30 April 2019; Accepted: 28 May 2019; Published: 29 May 2019

Abstract: We demonstrate a thermal tactile sensation display that can present various thermal sensations, namely cold/cool/warm/hot feelings, by varying the effective thermal conductivity of the display. Thermal sensation is one of the major tactile sensations and needs to be further investigated for advanced virtual reality/augmented reality (VR/AR) systems. Conventional thermal sensation displays present hot/cold sensations by changing the temperature of the display surface, whereas the proposed display is the first one that controls its effective thermal conductivity. The device contains an air cavity and liquid metal that have low and high thermal conductivity, respectively. When the liquid metal is introduced to fill up the air cavity, the apparent thermal conductivity of the device increases. This difference in the thermal conductivity leads to the users experiencing different thermal tactile sensations. Using this device, the threshold to discriminate the effective thermal conductivity was experimentally deduced for the first time. This thermal tactile display can be a good platform for further study of thermal tactile sensation.

Keywords: tactile display; thermal tactile display; thermal sensation; thermal conductivity; liquid metal

---

1. Introduction

Tactile displays have been studied to present pseudo-tactile sensations to users for advanced information communication technologies, such as efficient teleoperation and virtual reality/augmented reality (VR/AR) [1–7]. Tactile sensations are categorized into five sensations: wetness, roughness, hardness, pain sensation, and thermal sensation [8,9]. Roughness can be represented by the surface geometry. An array of micro-actuators can form various surface geometries, and therefore, many tactile devices to present various roughness sensation have been proposed, where microelectromechanical systems (MEMS) technologies have played an important role [10–12]. Hardness in tactile research is stiffness, to be precise. Stiffness is a material property and is difficult to control. In previous work, a magnetorheological fluid was encapsulated inside flexible membranes, whose apparent stiffness could be varied with the external magnetic field [13–15]. The encapsulation process was developed, and the device was used as the stiffness distribution display.

Thermal sensation is the sensation of cold/cool/warm/hot nature when we touch the surface of objects and plays a crucial role in tactile perception [16,17]. Conventionally, the thermal sensation displays present hot/cold surfaces by varying the surface temperature using Peltier elements [18–20]. The thermal module using the Peltier elements can be combined with other types of tactile displays, such as mechano-tactile display with an array of actuators [21,22] and electrostatic tactile display [23]. Thermal sensation display using radiation was proposed, which can stimulate subjects who are not in direct contact with the display [24]. Note that the thermal sensation is not determined only by the surface temperature, but also the thermal conductivity of the objects in contact [25]. For example, when we touch objects made of wood and metal which are at the same temperature, we perceive the metal to be colder than the wood, as illustrated in Figure 1. Materials with high thermal conductivity, such as metal, absorb heat from our finger, which leads to a cold/cool sensation. Control of thermal conductivity is challenging since it is an inherent physical property to each material and is determined by molecular configuration. Therefore, there have been no effective thermal tactile displays reported to date that can control their thermal conductivity.

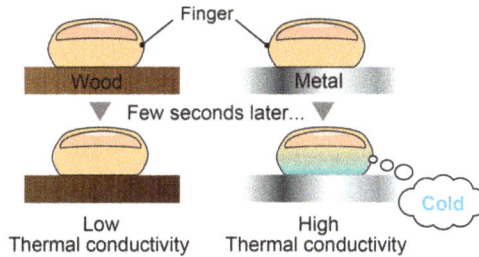

Figure 1. Illustration of heat transfer when we touch wood and metal at the same temperature with fingers.

In this work, we demonstrate a thermal tactile sensation display that can vary its thermal conductivity in a wide range and in an analogue manner by controlling its effective thermal conductivity. Effective thermal conductivity is the total thermal conductivity of the material and device, which depends not only on the material property, but also on the heterogeneous geometry. Since the temperatures of the device surface and the finger are room temperature (~25 °C) and body temperature (~37 °C), respectively, and are both low, we considered that the heat conduction through the device surface is dominant in the heat transfer. The conceptual sketch of the proposed tactile display is shown in Figure 2. The participant touches the display surface, which is a titanium plate. The display has an air cavity beneath the plate. Since the thermal conductivity of the air is as low as 0.024 W/mK at 0 °C, the effective thermal conductivity is also low at the original state. The display contains liquid metal encapsulated inside a flexible membrane. Liquid metal has equally high thermal conductivity to solid metals. As the liquid metal is supplied, it occupies the air cavity and the contact area between the liquid metal and the top plate increases. This increases the effective thermal conductivity of the display. The thermal property of this device was experimentally characterized and then perception tests were conducted to verify the effectiveness of this thermal tactile display. The property of thermal tactile perception was successfully characterized in a quantitative manner using the display. All the experimental protocols were approved by the Research Ethics Committee of Faculty of Science and Technology, Keio University.

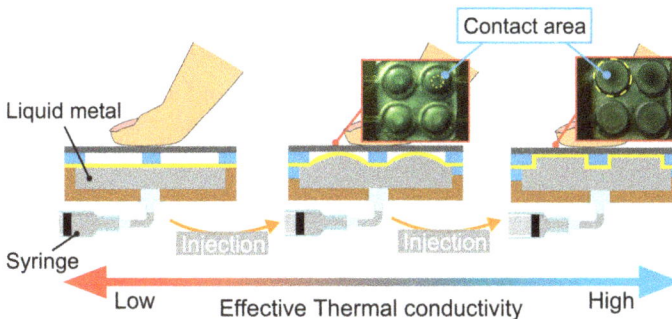

Figure 2. Conceptual sketch of the proposed thermal tactile display.

2. Principle and Fabrication Process

2.1. Principle

The working principle and the structure of the device are shown in Figures 2 and 3, respectively. The device encapsulates liquid metal with high thermal conductivity inside copper structures. The liquid metal we used for this device is Galinstan (68.5% gallium, 21.5% indium, and 10% tin), which has the thermal conductivity of 82 W/mK. It is sealed in the device with a latex rubber membrane with thermal conductivity of 0.13 W/mK. By controlling the amount of the liquid metal contacting the device's surface, different thermal conductivities can be presented to a fingertip which is to be placed on the top surface. As shown in Figure 2, at the initial state, liquid metal is separated from the titanium surface, resulting in a low thermal conductivity, which is determined by the geometry of the copper structures. When the liquid metal is injected into the device from the syringe underneath, it expands spherically with the latex rubber and reaches the titanium surface. Here, the thermal conductivity of the latex rubber is assumed to be negligible since its thickness is as thin as 20 μm. The larger the contact area it becomes, the better the effective thermal conductivity the device possesses. The effective thermal conductivity can be increased until the whole surface is fully in contact with the liquid metal. The contact area and thus the effective thermal conductivity can be controlled in the range continuously.

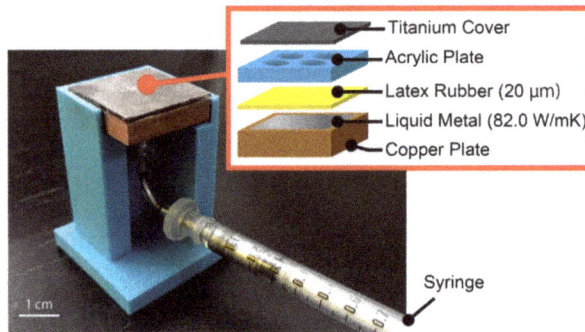

Figure 3. Structure of the thermal sensation tactile display.

2.2. Fabrication Process

Figure 4 shows the fabrication process of the thermal sensation tactile display. (a) Latex rubber is spin-coated onto a glass substrate at 1500 rpm for 30 s, which results in a membrane of 20 μm in thickness. Subsequently, the latex rubber was baked at 100 °C for 72 h. (b) An acrylic plate (20 mm × 20 mm × 1 mm) and a copper plate (20 mm × 20 mm × 5 mm) are processed with a numerical control (NC) cutting machine (MM-100, Modia Systems Co., Saitama Japan). Holes of 7 mm in diameter are drilled in a 2 × 2 array with intervals of 1 mm in the acrylic plate. In addition, two grooves with the size of 23 mm × 1 mm × 0.2 mm are formed on the acrylic plate, which pass through the center of the circles and work as the air escape paths when the air cavity decreases. Cavities of 17 mm × 17 mm × 2 mm and holes with diameters of 4.1 mm and 3 mm are processed onto a copper plate. (c) Both the acrylic plate and the copper plate are bonded to the latex rubber. (d) Liquid metal, Galinstan (Cool laboratory, Magdeburg, Germany), is injected into the cavity of the copper plate with a syringe. The syringe is connected to a luer fitting and a silicon tube so that the device can be set in a device holder. The cavity is sealed with polydimethylsiloxane (PDMS; Silpot 184 W/C, Dow Corning Toray Co., Ltd., Tokyo, Japan). (e) Finally, the titanium cover (20 mm × 20 mm × 0.05 mm) is bonded onto the acrylic plate of the device.

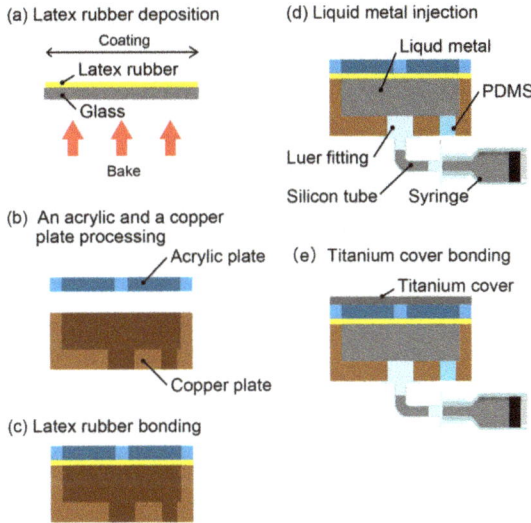

Figure 4. Fabrication process of the thermal sensation tactile display.

3. Experimental Procedure

3.1. Measurement of the Contact Area

The relationship between the amount of injected liquid metal and the contact area of the liquid metal and the titanic plate was investigated with a microscope. In this experiment, to visualize the contact area, the titanium cover was replaced by a glass plate.

3.2. Measurement of Effective Thermal Conductivity

The effective thermal conductivity of the device was measured with respect to the contact area using the flat plate comparison method (Figure 5) [26]. The effective thermal conductivity of the device was calculated using Fourier's law, as described in Equation (1).

$$Q = -A\lambda_1 \frac{(T_1 - T_2)}{l_1} = -A\lambda_2 \frac{(T_2 - T_3)}{l_2} \tag{1}$$

where Q is the heat transferred via the metal plate and the device. Since these materials are thermally insulated, the heat transfer quantity is constant. A is the contact area of the two materials. λ_1 is the thermal conductivity of a metal plate ($\lambda_1 = 83.5$ W/mK), and λ_2 is the thermal conductivity of the device. T_1, T_2, and T_3 are the measured temperatures at the interface between the metal plate and the hot plate, at the top surface of the device, and at the bottom of the device, respectively. l_1 and l_2 are the thicknesses of the metal plate ($l_1 = 5$ mm) and the device ($l_2 = 6$ mm), respectively. Note that all the other experiments were conducted at room temperature without the hot plate and the cool air.

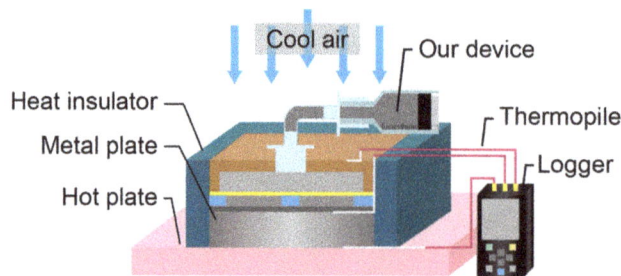

Figure 5. Experimental setup for measurement of thermal conductivity.

3.3. Evaluation of Thermal Sensation

A sensory experiment was carried out with ten participants (21 to 24 years old, 8 males and 2 females) to investigate how the thermal conductivity of the device affects the thermal sensation. First, the participants were requested to touch the device when the injection amount of the liquid metal was 0.00 mL and 0.08 mL, i.e., when the hottest and coldest sensation were supposed to be presented. Then, the participants were requested to touch the device with the injection amount ranging from 0.00 mL to 0.08 mL and score the coldness on a seven-item scale from 1 (cold) to 7 (hot).

3.4. Perceptual Threshold of Thermal Conductivity

Perception tests were conducted to deduce the perceptual threshold of thermal conductivity, if any, using the two-point identification method [27]. In this experiment, the injection amount was accurately controlled with a micro-syringe pump. We prepared two devices with different injection amount and thermal conductivity. The thermal conductivity of one device was fixed to be 75 W/mK, and the thermal conductivity of the other device was controlled between 80 W/mK and 105 W/mK with a step size of 5 W/mK in a random sequence. The participants were asked to touch the center of the device with the index fingers of both hands simultaneously, and then were asked whether they recognized a difference in coldness. To deduce the threshold more accurately, we conducted the same experiments varying the thermal conductivity with a step of 1 W/mK near the threshold. We consider that the initial temperatures of the finger and the device surface need to be consistent in the experiment. In the case that the participant uses his one finger to compare two conditions, a sufficiently long interval is needed between the perception tests. This interval may affect the answers, including reducing the accuracy. Therefore, we decided to request the participants to use both hands to detect the differences of thermal perception between the two conditions. The experimental conditions were as follows: contact time to the device—5 s; room temperature—25 °C.

4. Results and Discussion

4.1. Measurement of the Contact Area

Figure 6 shows the relationship between the amount of injected liquid metal and the contact area between the top surface and the liquid metal via a latex rubber membrane. The horizontal axis shows the injection amount of the liquid metal, and the vertical axis shows the contact area between the liquid metal and the titanium cover. The result shows that the contact area of the liquid metal increased with the injection amount in an almost linear manner.

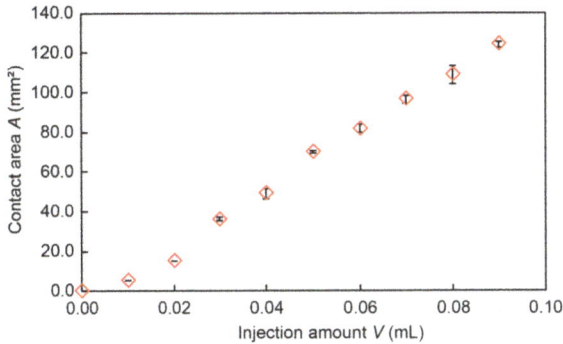

Figure 6. Measurement of the contact area.

4.2. Measurement of Thermal Conductivity

Figure 7 shows the relationship between the contact area and the thermal conductivity. The horizontal axis shows the contact area, and the vertical axis shows the measured thermal conductivity. The result indicated that the device could successfully present a wide range of thermal conductivities from 70.3 W/mK to 105 W/mK in a linear manner.

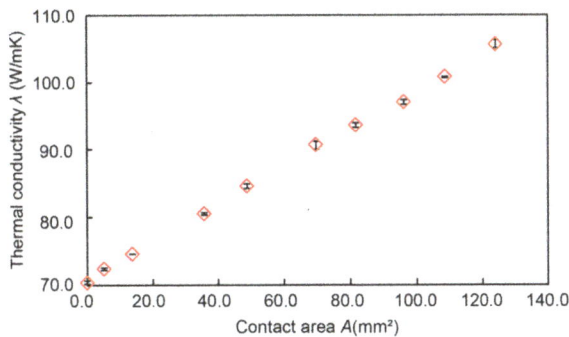

Figure 7. Measurement of thermal conductivity.

4.3. Evaluation of Thermal Sensation

Figure 8 shows the relationship between the thermal conductivity (horizontal axis) and the average and standard deviation of the scores of the thermal sensation (vertical axis). As the thermal conductivity increased, the scores of the thermal sensation increased, i.e., the participants felt the surfaces cooler. Significant differences were found between the initial state (70.3 W/mK) and 90.7 W/mK. These results verified the effectiveness of the proposed thermal tactile display. Though the case at the thermal conductivity of 90.7 W/mK showed significant difference against the one at 70.3 W/mK, the cases of 93.6 W/mK and 97.1 W/mK did not. All the cases have rather large standard deviation. Ambiguity in the thermal sensation needs to be taken into consideration when designing the thermal sensation display.

The human sensory experiment was conducted at room temperature, and the differences between the body temperature and the room temperature lead to change in the surface temperature of the device. One may consider that the participants detected the thermal sensation based on the surface temperature. However, if this is the case, the device with higher thermal conductivity should represent a warmer sensation, which is contrary to the experimental results. We conclude that in the experiments, the thermal conductivity was the dominant factor in determining the thermal sensation of the participants.

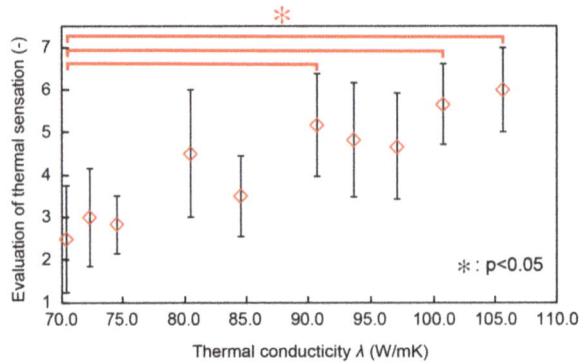

Figure 8. Evaluation of thermal sensation using the semantic differential (SD) method.

4.4. Perceptual Thershold on Thermal Conductivity

According to Figures 6 and 7, the standard deviation of the contact area and the thermal conductivity is no more than 4.6 mm^2 and 0.64 W/mK, respectively. From this result, we can say that the thermal conductivity could be controlled precisely by the injection amount of the liquid metal. In the experiments, we prepared two devices for both hands, which had the identical thermal conductivity with the same injection amount of liquid metal. Tables 1 and 2 show the analysis results of the experiment. The binomial test was used as an analysis method, for which the significance level was 5%. As a result of increasing the thermal conductivity by 5 W/mK (Table 1), when the difference of thermal conductivity between the two devices was 20 W/mK, a difference in the thermal sensation could be perceived. Then, the thermal conductivity was increased by 1 W/mK in the range of 91 W/mK to 94 W/mK. According to Table 2, when the difference of thermal conductivity was 18 W/mK, a difference in thermal sensation could be perceived. Therefore, it can be said that the perception threshold of thermal conductivity is approximately 18 W/mK. However, the perceptual threshold of thermal conductivity might not be the same in other ranges. Therefore, as a future study, we are planning to investigate the perception threshold within different ranges of thermal conductivity by changing the structure of the device. To the best of our knowledge, this is the first time that the thermal perception threshold has been experimentally deduced.

Table 1. Result of threshold experiment by using significant different judgement.

Thermal Conductivity (W/mK)	Value Difference (W/mK)	Significant Difference
75 vs. 80	5	NS
75 vs. 85	10	NS
75 vs. 90	15	NS
75 vs. 95	20	*
75 vs. 100	25	*
75 vs. 105	30	***

NS: Not significant; *** $p < 0.001$; * $p < 0.05$.

Table 2. Result of threshold experiment by using significant different judgement.

Thermal Conductivity (W/mK)	Value Difference (W/mK)	Significant Difference
75 vs. 91	16	NS
75 vs. 92	17	NS
75 vs. 93	18	*
75 vs. 94	19	*

NS: Not significant; * $p < 0.05$.

5. Conclusions

We successfully developed a thermal sensation display with controllable thermal conductivity. The encapsulated liquid metal increased the thermal conductivity as it contacts the device's top surface in a larger area. Control of the effective thermal conductivity with the contact area was successfully demonstrated. Using this device, the threshold of the thermal conductivity necessary to perceive differences in thermal sensation was experimentally deduced to be 18 W/mK. The proposed thermal sensation display can be readily applicable to the presentation of thermal sensations, i.e., hot/warm/cool/cold, for virtual/augmented reality applications and as a research platform for human thermal sensation.

Author Contributions: S.H. and N.M. designed and manufactured this display. S.H. and N.M. designed the experiments. S.H. analyzed the data. S.H. and N.M. wrote the paper.

Funding: This work was supported by a Grant-in-Aid for Scientific Research on Innovative Areas, Grant No. 18H05013.

Conflicts of Interest: The authors declare no conflict of interest.

References

1. Storonks, C.H.; Parker, J.D.; Stacey, A.; Barnes, N. Psychophysical Evaluation of a Tactile Display Based on Coin Motors. *Artif. Organs.* **2018**, *8*, 1224–1233. [CrossRef] [PubMed]
2. Gibson, A.; Webb, A.; Stirling, L. Evaluation of a Visual-Tactile Multimodal Display for Surface Obstacle Avoidance During Walking. *IEEE Trans. Human-Mach. Syst.* **2018**, *48*, 604–613. [CrossRef]
3. Hansen, V.K.; Eskildsen, S.M.; Larsen, V.O. Region-of-Interest based Finite Element Modelling of the Brain—An approach to Brain Surgery Simulation. *Proc. Fourteenth Int. Conf. Pattern Recognit.* **1998**, *1*, 292–296.
4. Ogawa, K.; Ibrahim, Y.; Ohnishi, K. Development of Flexible Haptic Forceps Based on the Electro-Hydraulic Transmission System. *IEEE Trans. Ind. Inform.* **2018**, *14*, 5256–5267. [CrossRef]
5. Liu, S.W.; Li, Y. The research for control strategies and methods of teleoperation system. *World Autom. Congr.* **2012**, *1*, 2–5.
6. Hayashi, K.; Tamura, T. Teleoperation Performance using Excavator with Tactile Feedback. *2009 IEEE Int. Conf. Mechatron. Autom. ICMA* **2009**, *5*, 2759–2764.
7. Kumazawa, I.; Yano, S.; Suzuki, S.; Ono, S. A multi-modal interactive tablet with tactile feedback, rear and lateral operation for maximum front screen visibility. *Proc. IEEE Virtual Real.* **2016**, *23*, 211–212.
8. Shinoda, H. Tactile Sensing for Dexterous Hand. *J. Robot. Soc.* **2000**, *18*, 767–771. [CrossRef]
9. Tachi, S. Telexistence: Enabling Humans to Be Virtually Ubiquitous. *IEEE Comput. Graph. Appl.* **2016**, *36*, 8–14. [CrossRef]
10. Kosemura, Y.; Hasagawa, S.; Miki, N. Surface properties that can be displayed by a microelectromechanical system-based mechanical tactile display. *Micro Nano Lett.* **2016**, *11*, 240–243. [CrossRef]
11. Ishizuka, H.; Miki, N. MEMS-based tactile displays. *Displays* **2015**, *37*, 25–32. [CrossRef]
12. Na, K.; Han, J.; Roh, D.; Chae, B.; Yoon, E.; Kang, Y.J.; Cho, I. Flexible latching-type tactile display system actuated by combination of electromagnetic and pneumatic forces. *Proc. IEEE Int. Conf. Micro Electr. Mech. Syst.* **2012**, *2*, 1149–1152.
13. Ishizuka, H.; Miki, N. Development of a tactile display with 5 mm resolution using an array of magnetorheological fluid. *Jpn. J. Appl. Phys.* **2017**, *56*, 06GN19. [CrossRef]
14. Ishizuka, H.; Miki, N. Fabrication of hemispherical liquid encapsulated structures based on droplet molding. *J. Micromech. Microeng.* **2015**, *25*, 125010. [CrossRef]
15. Ishizuka, H.; Rorenzoni, N.; Miki, N. Tactile display for presenting stiffness distribution using magnetorheological fluid. *Mech. Eng. J.* **2014**, *1*, 1–10. [CrossRef]
16. Kobayashi, T.; Fukumori, M. Proposal of a Design Tool for Tactile Graphics with Thermal Sensation. In Proceedings of the 2012 18th International Conference on Virtual Systems and Multimedia, Milan, Italy, 2–5 September 2012; pp. 537–540.

17. Yamamoto, A.; Cros, B.; Hashimoto, H.; Higuchi, T. Control of Thermal Tactile Display Based on Prediction of Contact Temperature. *Int. Conf. Robot. Autom.* **2014**, *2*, 1536–1541.
18. Yang, G.; Kyung, K.; Srinivasan, A.M.; Kwon, D. Quantitative Tactile Display Device with Pin-Array Type Tactile Feedback and Thermal Feedback. *Proc. IEEE Int. Conf. Robot. Autom.* **2006**, *1*, 3917–3922.
19. Mao, Z.; Wu, J.; Li, J.; Zhou, L.; Li, X.; Yang, Y. A Thermal Tactile Display Device with Multiple Heat Sources. *Proc. 2012 Int. Conf. Ind. Control Electron. Eng. ICICEE 2012* **2012**, *1*, 192–195.
20. Ino, S.; Shimizu, S.; Odagawa, T.; Sato, M.; Takahashi, M.; Izumi, T.; Ifukube, T. A Tactile Display for Presenting Quality of Materials by Changing the Temperature of Skin Surface. *Work. Robot Hum. Commun.* **1993**, *1*, 220–224.
21. Yang, G.; Kyung, K.; Srinivasan, A.M.; Kwon, D. Development of Quantitative Tactile Display Device to Provide Both Pin- Array-Type Tactile Feedback and Thermal Feedback. In Proceedings of the Second Joint EuroHaptics Conference and Symposium on Haptic Interfaces for Virtual Environment and Teleoperator Systems (WHC'07), Tsukaba, Japan, 22–24 March 2007; pp. 578–579.
22. Yang, T.; Yang, G.; Kwon, D.; Kang, S. Implementing Compact Tactile Display for Fingertips with Multiple Vibrotactile Actuator and Thermoelectric Module. *Proc. IEEE Int. Conf. Robot. Autom.* **2007**, *1*, 7–8.
23. Ishizuka, H.; Hatada, R.; Cortes, C.; Miki, N. Development of a fully flexible sheet-type tactile display based on electrovibration stimulus. *Micromachines* **2018**, *9*, 230. [CrossRef]
24. Saga, S. Thermal-Radiation-Based Haptic Interface. *Haptic Interact.* **2015**, *277*, 105–107.
25. Shen, H.; Jiang, S.; Sukigara, S. Dependence of Thermal Contact Properties on Compression Pressure. *J. Fiber Sci. Technol.* **2017**, *73*, 177–181. [CrossRef]
26. Zhao, D.; Qian, X.; Gu, X.; Jajja, A.S.; Yang, R. Measurement Techniques for Thermal Conductivity and Interfacial Thermal Conductance of Bulk and Thin Film Materials. *J. Electron. Packag.* **2016**, *138*, 1–19. [CrossRef]
27. Tong, J.; Mao, O.; Goldreich, D. Two-point orientation discrimination versus the traditional two-point test for tactile spatial acuity assessment. *Front. Human Neurosci.* **2013**, *7*, 1–11. [CrossRef] [PubMed]

micromachines

MDPI

Article

Separation of Nano- and Microparticle Flows Using Thermophoresis in Branched Microfluidic Channels

Tetsuro Tsuji, Yuki Matsumoto, Ryo Kugimiya, Kentaro Doi and Satoyuki Kawano *

Graduate School of Engineering Science, Osaka University, Toyonaka, Osaka 560-8531, Japan;
tsuji@me.es.osaka-u.ac.jp (T.T.); ymatsumoto@bnf.me.es.osaka-u.ac.jp (Y.M.); zeath.00@gmail.com (R.K.);
doi@me.es.osaka-u.ac.jp (K.D.)
* Correspondence: kawano@me.es.osaka-u.ac.jp; Tel.: +81-6-6850-6175

Received: 26 April 2019; Accepted: 7 May 2019; Published: 12 May 2019

Abstract: Particle flow separation is a useful technique in lab-on-a-chip applications to selectively transport dispersed phases to a desired branch in microfluidic devices. The present study aims to demonstrate both nano- and microparticle flow separation using microscale thermophoresis at a Y-shaped branch in microfluidic channels. Microscale thermophoresis is the transport of tiny particles induced by a temperature gradient in fluids where the temperature variation is localized in the region of micrometer order. Localized temperature increases near the branch are achieved using the Joule heat from a thin-film micro electrode embedded in the bottom wall of the microfluidic channel. The inlet flow of the particle dispersion is divided into two outlet flows which are controlled to possess the same flow rates at the symmetric branches. The particle flow into one of the outlets is blocked by microscale thermophoresis since the particles are repelled from the hot region in the experimental conditions used here. As a result, only the solvent at one of outlets and the residual particle dispersion at the other outlet are obtained, i.e., the separation of particles flows is achieved. A simple model to explain the dynamic behavior of the nanoparticle distribution near the electrode is proposed, and a qualitative agreement with the experimental results is obtained. The proposed method can be easily combined with standard microfluidic devices and is expected to facilitate the development of novel particle separation and filtration technologies.

Keywords: microscale thermophoresis; multiphase flow; microfluidic channels; nano/microparticle separation; micro-electro-mechanical-systems (MEMS) technologies

---

1. Introduction

Fluids that contain dispersed phases, such as nano- and microparticles, appear in a wide range of applications. For instance, a nanofluid is a class of fluid that contains nanometer-sized materials, in which a significant increase in the heat-transfer rate compared to conventional engineered fluid has been reported [1,2]. Ever since its discovery, the nanofluid has served in several engineering applications, e.g., fuel-cells [3–5], porous materials [6–8], and petroleum engineering applications [9–12]. In addition to these applications, nanofluids have become an important research topic in the development of lab-on-a-chip (LOC) devices, where the sorting and/or accumulation of target nanomaterials in fluidic devices are necessary for a controlled chemical reaction or an analysis of the targets [13,14]. In LOC devices, controlling and separating the flow of tiny dispersed phases, such as nanoparticles in nanofluids, is one of the main concerns in the field of microfluidics and nanofluidics [15–21]. To be more specific, flow control of nanoparticles can enhance the detection/identification performance of the biosensor [22] installed in a LOC device. Therefore, various techniques to control the particle flow in microfluidic channels have been proposed in [15–20]. For instance, electrophoresis [23,24] and dielectrophoresis [25] in micro- and nanofluidic devices are widely acknowledged experimental techniques using electrokinetic effects.

However, applying electrical potential differences across the device may result in electroosmotic flows and/or the electrolysis of solvents, which may complicate data analysis and be undesirable depending on a situation. Other than these electrokinetic methods, other driving mechanisms such as diffusiophoresis [26–30] and thermophoresis [31–33] have also been actively investigated recently. Since each method has a different physical basis, one may choose a suitable technique according to the properties of the target materials.

Among these driving mechanisms, thermophoresis is expected to introduce a new direction for particle flow control since it is sensitive to the composition of a continuous phase and the nature of particle characteristics, such as chemical surface modifications. More specifically, the direction of particle motion may be controlled by choosing a proper experimental setting. For instance, it has been shown that the direction of thermophoresis can be reversed by controlling the average temperature of the solution [34–36] or by adding electrolytes [37–40] or polymers [41,42] into the solution. Owing to the high sensitivity of thermophoresis, depending on the nature of particle and/or solvent, microscale thermophoresis has recently been developed to evaluate protein-binding [43,44]. Here, the prefix terminology "microscale" is used to emphasize that the driving temperature distribution is localized in a microscale spatial region. However, although there have been theoretical [33,45–48] and simulation-based [49–52] approaches to understand the nature of thermophoresis, its physical mechanism is not yet fully understood.

Given such a growing interest in thermophoretic manipulation, the previous study by the authors tried to apply microscale thermophoresis to particle flow control in LOC devices [53,54]. More specifically, using micro-electro-mechanical-systems (MEMS) technologies, a micro heater was installed in a straight microfluidic channel, and the on-chip thermophoretic separation device was developed [53]. The counterbalance between the flow of the continuous phase and the thermophoresis of microparticles resulted in the formation of a localized particle distribution in the straight channel. However, the particle flow separation to branched channels, as demonstrated in [15–17,19,20], was not achieved in the previous study. It is useful in LOC applications to transport dispersed phases to a desired branch. The present paper is an extension of the previous study [53] on branched microfluidic channels, which are more suitable for the separation of a dispersed phase from a continuous one. To eliminate unnecessary complexity, a symmetric Y-shaped branch is used as a microfluidic channel. In this way, the effect of thermophoresis on the particle flow separation is elucidated. Moreover, a simple numerical model is introduced to explain the separation dynamics at the Y-shaped branch. In the present study, the size of the particle is reduced from microscale to nanoscale to demonstrate that the present approach is also applicable not only to cells with $O(1)$ μm but also to viruses [55,56] or pollen allergen particles [57] with $O(10)$–$O(100)$ nm, broadening the scope of application of state-of-the-art micro- and nanofluidics in biosciences. The present demonstration of nano- and microparticle flow separation using thermophoresis suggests the function of selective particle flow control may feasibly be installed to existing microfluidic devices.

2. Experimental Methods

2.1. Details of Microfluidic Devices

Microfluidic channels are fabricated by bonding a block made of polydimethylsiloxane (PDMS) and a glass substrate using a similar protocol to that described in the previous study [53]. In the present paper, the PDMS block has a Y-shaped branch in a microfluidic channel. The schematic of the test section near the branch is shown in Figure 1a. The channel has a uniform cross-section with a height $h = 17.2$ μm and a width $w = 450$ μm. The dimension of the cross-section of the channel is similar to that in the previous study [53], where unwanted thermal convection was confirmed to be absent thanks to the small channel height. Three holes with a diameter 2 mm for an inlet and two outlets are fabricated, as shown in the inset of Figure 1a. These holes are connected to reservoirs using silicone tubes, as shown in Figure 1b. The distance from the holes to the Y-shaped branch, i.e., the channel lengths, is $L = 5$ mm for the inlet and two outlets. This technique is considered in terms of the fluid dynamics of the continuous phase to realize a fully developed flow at the test section and almost equal flow rates in two branched outlets. To induce a temperature increase for thermophoresis of the dispersed phase, an electrode heater, which has the boundary condition of uniform heat flux, is used, as shown in Figure 1a, where the electrode width is $w_{elec} = 20$ µm and thickness is 150 nm. The electrode thickness is thin compared to the channel height h, and it does not affect the flows. The fabrication process is described in the following.

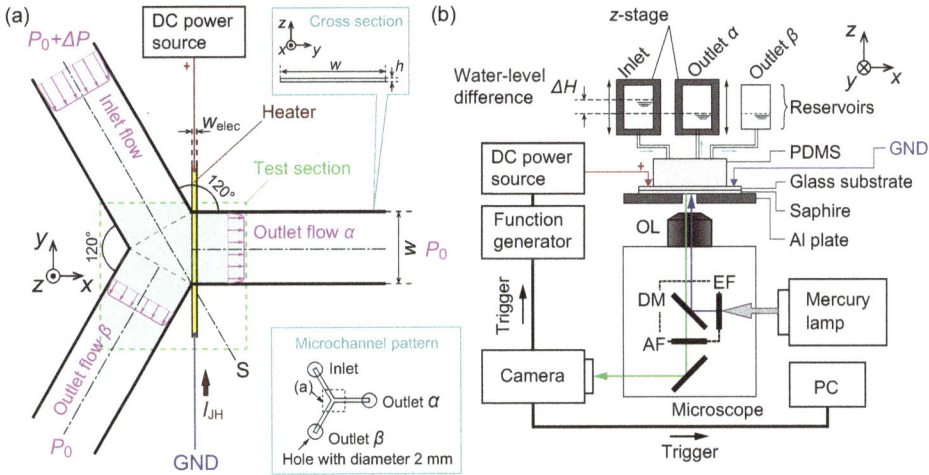

Figure 1. (**a**) Schematic of the test section. The branched microfluidic channel has a rectangular cross-section in the yz plane with a height $h = 17.2$ µm and a width $w = 450$ µm. The inlet flow is divided into two outlet flows α and β. A thin-film electrode heater is fabricated at the entrance of the outlet flow α. Flow profiles of the inlet and outlets are schematically drawn based on the analytical solution of the Poiseuille flow in a rectangular channel [58]. (**b**) Overview of the experimental setup. EF: emission filter. AF: absorption filter. DM: dichroic mirror. OL: objective lens. PC: personal computer.

2.1.1. Fabrication of a PDMS Block

A mold for the microfluidic channel pattern is prepared on a Si substrate by a photolithography of negative photoresist SU-8 3005 (MicroChem Corp., Westborough, MA, USA). The PDMS block is cast from the mold to obtain the microfluidic channel pattern. The height of the mold, which determines the height of the microfluidic channel h, is measured as $h = 17.2 \pm 0.2$ µm by scanning the PDMS block using a laser displacement sensor (LK-H008W, Keyence, Osaka, Japan).

2.1.2. Fabrication of the Electrode Pattern on the Glass Substrate

The glass substrate is sonicated in dimethylformamide (DMF), ethanol, and ultra-pure water in series for 15 min each. After drying out the glass substrate at 200 °C for 5 min, an Au thin-film is deposited on the substrate by sputtering, where the thickness of the Au layer is 150 nm. Here, Cr is used as an adhesion layer between the glass and Au. A positive photoresist (AZ5214E, Merck, Germany) is spin-coated onto the substrate. After a prebake at 90 °C for 2 min, the substrate is exposed to UV light with 20 mJ·cm^{-2} through a photomask to obtain an electrode pattern. The substrate is then immersed in a developer solution (a mixture of 1:1 ultra-pure water and AZ developer; Merck, Germany) to obtain the photoresist layer with the electrode pattern. After the postbake at 120 °C for 2 min, the substrate is immersed into etching solutions for Au (AURUM-301, Kanto Chemical Co., Inc., Tokyo, Japan) and Cr (Mixed-Acid Cr Etching solution, Kanto Chemical Co., Inc., Tokyo, Japan) layers. Finally, the substrate is sonicated in acetone and ultra-pure water.

The electrode heater has an electrical resistance of $23.3 \pm 1.7 \, \Omega$, i.e., the variation in the fabrication error is 7%. The error may be attributed to the presence of a gap between the photomask and the substrate during UV exposure, which determines the accuracy of the pattern transfer; this is difficult to control precisely using a manual mask aligner such as that which is used in the present fabrication. For the present research, the error is within the acceptable range, but a direct pattern exposure system, such as a laser lithography system, will be necessary to achieve further miniaturization of the electrode.

2.1.3. Bonding Process

The contact surfaces of the PDMS block and the glass substrate are treated by oxygen plasma (RIE-10NR, Samco, Kyoto, Japan) to enhance the adhesion. The bonding process is carried out using an aligner so that the electrode is placed at the entrance of the outlet α, as shown in Figure 1a. A direct current (DC) power source (PAN35-10A, Kikusui Electronics Corp., Yokohama, Japan) is connected to the electrode. An electric current I_{JH} produces the Joule heat from the electrode and a microscale temperature distribution is formed at the entrance of the outlet α. In the previous study [53], where the dimensions of the electrode were the same as that of the present study, the maximum temperature T_{max} near the electrode was measured to be about 360 K. Therefore, a similar temperature increase is expected in the present device. Finally, the inlet and outlets are connected to the reservoirs by silicone tubes.

2.2. Experimental Setup

The complete experimental setup is shown in Figure 1b. An inverted microscope (IX-71, Olympus, Tokyo, Japan) with an objective lens (OL, 10x magnification, numerical aperture = 0.3) and a scientific complementary metal-oxide-semiconductor (sCMOS) camera (Zyla 5.5, Andor Technology Ltd., Tokyo, Japan) are used for observation of the device. To prevent the overall temperature increase of the device, it is placed on sapphire glass, which has high thermal conductivity and optical transmissivity. The DC power source is controlled by a function generator (WF1973, NF, Kanagawa, Japan). A trigger signal from the camera synchronizes the image acquisition in a personal computer (PC) and the onset of Joule heating through the function generator.

A mercury lamp (U-HGLGPS, Olympus, Tokyo, Japan) is used as the illumination light source. The illumination light goes through an excitation filter (EF) and is converted to the excitation light. Being irradiated by the excitation light, micro- or nanoparticles in the device emit fluorescence, which is monitored by the camera through an absorption filter (AF).

The flow rate within the device is controlled by water-level differences between the reservoirs. First, the z-stage, which holds the reservoir for outlet α, is manipulated to eliminate the water-level difference between the reservoirs for outlets α and β. Then, the particle flow becomes symmetric with respect to a plane, S, shown in Figure 1a. Next, the z-stage, which holds the reservoir for the inlet, is manipulated to stop the flow in the microfluidic channel, i.e., all the water-levels in three reservoirs are controlled to be the same. Then, the reservoir for the inlet is lifted by ΔH, as shown in Figure 1b, to induce the fluid flow of a sample solution with a required flow rate. As discussed in [54], the generated pressure difference ΔP in Figure 1 is estimated as $\Delta P = \rho g \Delta H$, where ρ is the mass density of the sample solution and $g = 9.8 \, \text{m·s}^{-2}$ is the acceleration of gravity. In this research, an aqueous solution is used and thus $\rho = 1.0 \times 10^3 \, \text{kg·m}^{-3}$. Because the resolution of the z-stage is 1 μm, the resolution of ΔP can be estimated as 1×10^{-2} Pa. The resulting flow fields will be discussed in Section 3.1.

2.3. Sample Solutions

Polystyrene (PS) particles are used as a dispersed phase. In the experiments, PS particles with carboxylate surface modifications in a Tris-HCl aqueous buffer (pH = 8.0, 321-90061, Nippon Gene Co., Ltd., Tokyo, Japan) are used since this choice was confirmed to yield thermophoresis in the previous experiments [53] when the particle diameter d was 0.99 ± 0.022 μm. The concentration of

Tris-HCl is 10 mM. To avoid the occurrence of inter-particle interactions, the concentration of particles should be dilute and is set to be less than 4×10^{-2} wt%. These sample solutions were prepared using ultra-pure water. Microparticle ($d = 0.99 \pm 0.022$ μm, F8823, Molecular Probes, Eugene, OR, USA) and nanoparticles ($d = 99 \pm 8$ nm, F8803, Molecular Probes, Eugene, OR, USA) are tested in the present paper.

2.4. Procedures

As described in Section 2.2, the mean flow in the microfluidic channel is induced, where the flow is symmetric with respect to the plane S in Figure 1a. At $t = 0$ s, heating the microfluidic channel is induced by applying the electric current $I_{JH} = 4 \times 10^{-2}$ A, and recording the subsequent behaviors of PS particles. The duration of the experiment is set to 300 s. To focus on the effect of temperature increase, it must be ensured that the pressure difference ΔP does not change during the experiments of 300 s. This is confirmed as follows. As will be shown in Section 3.1, the flow speed in the inlet channel is less than 10 μm·s^{-1}. That is, the flow rate, which is obtained by multiplying the flow speed by the cross-sectional area $wh = 7.7 \times 10^3$ μm^2, is estimated as 7.7×10^{-14} m^3·s^{-1}. Due to mass conservation, this flow rate must be compensated by the decrease (and increase) of the reservoir water level in the inlet (and outlets). The cross-section of the reservoir is 2.3×10^{-4} m^2, that is, ΔH decreases with the speed 3.3×10^{-10} m·s^{-1} to compensate for the mass flow in the microfluidic channel. For the experiment of 300 s, the difference between $\Delta H(t = 0$ s$)$ and $\Delta H(t = 300$ s$)$ is estimated as, at most, 1.0×10^{-7} m, which corresponds to $\Delta P(t = 0$ s$) - \Delta P(t = 300$ s$) = \rho g[\Delta H(t = 0$ s$) - \Delta H(t = 300$ s$)] \approx O(10^{-3})$ Pa. Since $\Delta P(t = 0$ s$) = 0.5$ or 1.0 Pa is used in the present paper, it is considered that the variation of ΔP is negligibly small in the experiments. In other words, the mean flow can be considered to be in a steady state during the experiments.

3. Results and Discussion

3.1. Flow Fields

Thermophoresis is a rather weak effect, that is, it can be hindered by the fast flow of continuous phase. Therefore, to observe the effect of thermophoresis effectively, we should induce a creeping flow of $O(1)$–$O(10)$ μm·s^{-1} in the microfluidic channel [53,54]. However, in general, the creeping flow is difficult to control since a finer pressure control resolution is required. In this section, the reliability of the control on the flow of the continuous phase is investigated.

The flow field in the absence of temperature increase is presented, where ΔP is set to 1.0 Pa. Figure 2a shows the experimental result obtained by the particle image velocimetry (PIV) using the microparticle as a tracer, where the inset shows corresponding stream lines. Since the channel height $h \approx 17$ μm is small, the fluorescence of particles in the entire z-direction is recorded by the camera, i.e., the PIV result is considered to be the average in the z-direction. The inlet flow is separated at the Y-shaped branch into two symmetric outlet flows. The flow speed is about 4 μm·s^{-1} and 2 μm·s^{-1} for the inlet and the outlets, respectively. To validate the experimental results, the flow field obtained from a numerical simulation using a finite element method is shown in Figure 2b. The simulation is carried out by a commercial software, COMSOL Multiphysics 5.3 (COMSOL, Inc., Stockholm, Sweden). The result shows the velocity vector u at a plane $z = h/2$. The overall flow profile is consistent with the experimental result in Figure 2a, although the magnitude of the flow is larger than that observed in the experiment. For the Poiseuille flow between two parallel plates, the flow speed averaged in the z-direction is 2/3-times the maximum at $z = h/2$. Therefore, multiplying the result in Figure 2b by a factor 2/3 is expected to yield the result in Figure 2a.

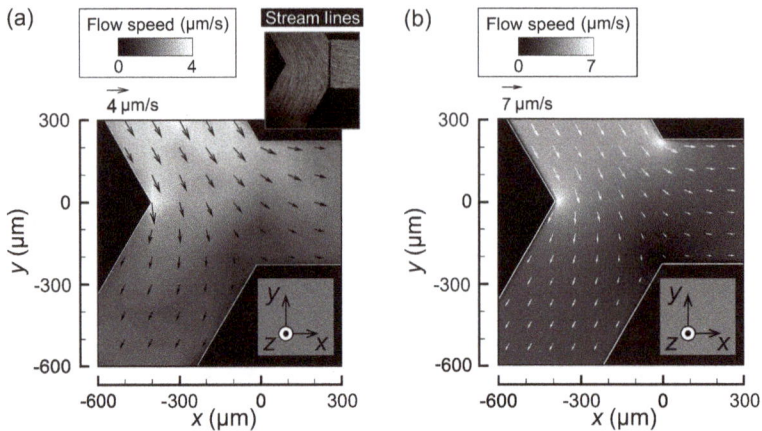

Figure 2. Flow field at the test section without Joule heating for $\Delta P = 1 \times 10^{-2}$ Pa. The inlet flow is equally separated into two outlet flows. (**a**) Experimental result obtained by the particle image velocimetry (PIV) analysis. (**b**) Numerical result at $z = h/2$ obtained by the simulation using a finite element method.

Furthermore, the simulation result is validated using the theoretically obtained inlet flow speed V_{in}, which is the flow speed averaged over the cross-section. The channel length is $L = 5$ mm, the experimentally observed flow speed is $V_{in} = O(1) - O(10)$ µm·s^{-1}, and the viscosity of the solution at room temperature is $\eta = 8.94 \times 10^{-4}$ Pa·s. Then, laminar flow in the microfluidic channel can be assumed, since the Reynolds number is estimated as $Re = \rho V_{in} L / \eta < 6 \times 10^{-2}$ and is sufficiently small. Neglecting the minor pressure losses, such as velocity head, entrance loss, and branch loss, the pressure difference $\Delta P = 1$ Pa should be compensated by the friction losses along the microfluidic channel. Considering the incompressible and Newtonian fluid with small Reynolds number, the hydraulic resistances of the inlet R_{in} and that for the outlet R_{out} are expressed as $R_{in} = \frac{12\eta L}{h^3 w}[1 - 0.630(h/w)]^{-1} (= R_{out})$, where the Poiseuille flow in a thin square duct is assumed [58,59]. Then, using Bernoulli's theorem, the relation $\Delta P = R_{in} Q_{in} + R_{out} Q_{out}$ holds, where Q_{in} and Q_{out} are the flow rates in the inlet and outlet, respectively. It should be noted that Q_{out} values in both outlets are equal due to the symmetric nature of the microfluidic channel. Then, due to the mass conservation, $Q_{in} = 2Q_{out}$ holds. Therefore, it is concluded that $\Delta P = (3/2)R_{in}Q_{in}$, which turns out to be $Q_{in} = (2/3)\Delta P/R_{in}$. The average flow speed $V_{in} = Q_{in}/(wh)$ given by the above relation is $V_{in} = 3.5$ µm·s^{-1}. This theoretical value overestimates the simulation value 3.14 µm·s^{-1} by 10%, where the simulation value is evaluated at the position $(x, y) = (-320$ µm, 280 µm$)$ in the inlet. The overestimation may be due to the neglected minor losses. Nonetheless, the agreement among the experiment, the simulation, and the theory is reasonable, and it is concluded that the control of the continuous phase is adequate to carry out the thermophoresis experiments.

3.2. Microparticle Flow Separation

In this section, the result for microparticles with $d = 1$ µm using the pressure difference $\Delta P = 1.0$ Pa is presented. As demonstrated in the previous work [53], the PS microparticles under the present experimental conditions were thermophobic, that is, the particles were repelled from the hot region. Therefore, it was expected that the particles would move away from the thin-film electrode.

Figure 3a shows the snapshot at $t = 0$ s, at which the Joule heating of the thin-film electrode starts. According to the previous study [53] using the same electrode heater, the temperature near the electrode gradually increases with increasing time. Since the continuous phase at room temperature is supplied from the inlet, a non-uniform temperature field is formed near the electrode. The temperature field reaches an almost steady state in several seconds, producing a magnitude of the temperature gradient of about 0.6 K·μm^{-1} near the electrode. It should be noted that the fluid flow is driven by the pressure difference from the inlet to outlets α and β. As time progresses, the particle flow separation begins to be observed, as shown in Figure 3b. More specifically, it seems that microparticles cannot enter the outlet α, and the particle-concentrated region emerges near the thin-film electrode upstream. As a result, the particles are flushed out in the outlet α, and the high-concentration dispersion is obtained in the outlet β. At $t = 250$ s, few particles exist in the outlet α, resulting in complete particle flow separation.

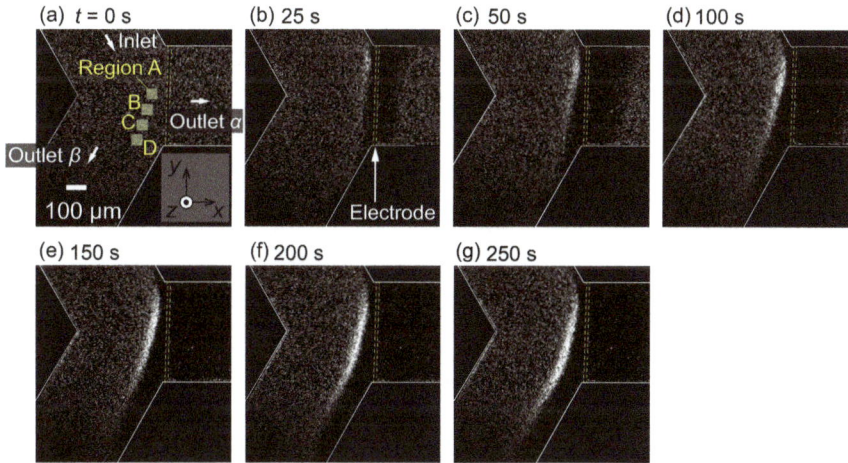

Figure 3. (**a–g**) Time series of the particle flow separation induced by microscale thermophoresis for the case with a particle diameter $d = 1$ μm and $\Delta P = 1.0$ Pa. At $t = 0$ s, the heating by electrode is initiated. Particle flow from the inlet is separated at the Y-shaped branch. Because the thermophoresis is directed to the colder region, the PS particles cannot enter the outlet α.

The increase in fluorescence of particular regions A, B, C, and D indicated in Figure 3a is evaluated to investigate the separation process in more detail. It should be noted that the increase in the fluorescence indicates the increase in the particle concentration. Figure 4 shows the time-development of the fluorescence intensity for these four regions. It is found that the intensity increases as time progresses for all regions; however, the onset of the increase in region A is earlier than that in region D. The time-dependent behaviors in Figure 4 are compared with the snapshots in Figure 3. At the early stage of the experiment, as shown in Figure 3b–d and Figure 4 for $t \leq 100$ s, the particles begin to accumulate in the region where the flow drag is counterbalanced by the thermophoretic force, as in [53]. At the later stage of the experiment, as shown in Figure 3d–g and 4 for $t \geq 100$ s, these concentrated particles are gradually transported to the outlet β as time progresses. Therefore, the increase in the fluorescence intensity in regions B, C, and D occurs later, as shown in Figure 4. The saturation of the intensity in region A and the intensity increase in region D are caused by the transport of these particles to the outlet β.

For the concentration range investigated in the present study, no apparent particle concentration effect is observed since the initial concentration of $O(10^{-2})$ wt% is sufficiently diluted to neglect the finite-size effect and the inter-particle interaction. In the case of the straight channel of the previous study [53], the concentration increased by 100-fold to 4 wt% after 5 min of operation with a similar flow rate. When the device is operated for a longer duration, such a large concentration ratio may cause the finite-size effect and/or the inter-particle interaction that complicate the phenomena. On the contrary, the saturated particle density observed in Figure 4 is more favorable than that in the straight channel since the present device can avoid reaching too high a particle concentration even with a longer operation time. That is to say, the branched channel is suitable for practical applications where continuous particle flow separation may be required. With the aim of achieving more effective particle separation with a larger flow rate, the design of both the flow and temperature fields are important because the position of the highly concentrated particle region and the value of the saturated particle density are determined as the result of the interaction between these two fields. The present demonstration is a first step toward designing better channel and electrode patterns.

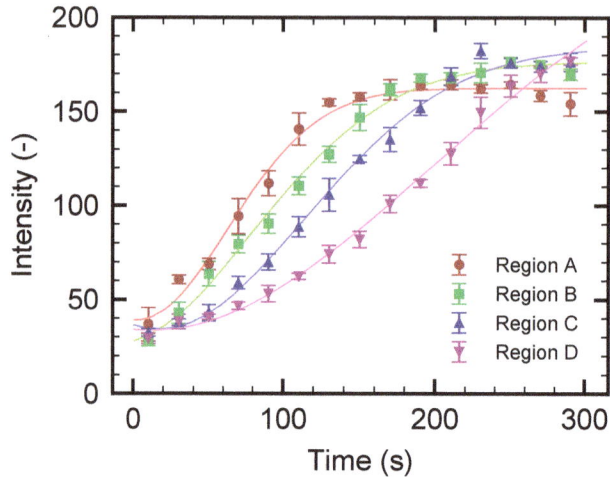

Figure 4. Time-development for fluorescence intensity of microparticles in the regions A, B, C, and D indicated in Figure 3a. The pressure difference ΔP is set to 1.0 Pa.

3.3. Nanoparticle Flow Separation

In this section, the results for nanoparticles with $d = 100$ nm are presented. In the previous study [54], it was demonstrated that the PS nanoparticles were repelled from the hot spot produced by laser irradiation. Although the sample solution in [54] contained an additional surfactant, it was expected that the nanoparticles in the present experimental condition were also repelled from the hot region since the addition of a surfactant did not result in a reversal of the thermophoresis direction of PS microparticles in [53].

First, $\Delta P = 1.0$ Pa is used as in Section 3.2. Figure 5 shows the snapshots of the obtained video images. It should be noted that the nanoparticles are too small to distinguish each of them with the present optical setup. This is the reason for using florescent nanoparticles. As time goes on, the left side of the electrode starts to become bright, indicating the increase of fluorescent nanoparticles. At $t \geq 200$ s, the lower-left region near the electrode, indicated by an arrow in Figure 5f, shows an apparent increase in the fluorescence intensity. This is clearly seen from Figure 5h, which is the magnification of a rectangular region indicated in Figure 5b. The increased nanoparticle concentration is then transported to the outlet β. This behavior of nanoparticles is qualitatively similar to that of microparticles presented in Figure 3; however, the position of the particle-concentrated region is closer to the electrode in this case. Such a difference is attributed to the difference in thermophoretic mobilities between micro- and nanoparticles. These results indicate that the nanoparticles are more insensitive to the temperature gradient and/or they have larger diffusion coefficients D than microparticles, as expected from the Stokes–Einstein relation $D = k_{\mathrm{B}}T/(3\pi\eta d)$, where k_{B} is the Boltzmann constant and T is the temperature. Figure 6 shows the time-development of the fluorescence intensity in regions A, B, C, and D indicated in Figure 5a. It is found that the intensity for all the regions increases with increasing time; however, the increment is smaller than that for the experiment using microparticles, as shown in Figure 4. More specifically, the intensity is, at most, three times higher than that at the initial state $t = 0$ s, indicating an almost three-fold increase in nanoparticle concentration. The increase of the intensity in region D, which is placed at the entrance to the outlet β, is slightly delayed but larger compared with that of other regions, because the concentrated nanoparticles in regions A, B, and C are transported to the outlet β as time progresses. The dynamic behavior of nanoparticles will be discussed in Section 3.4 using a simple model.

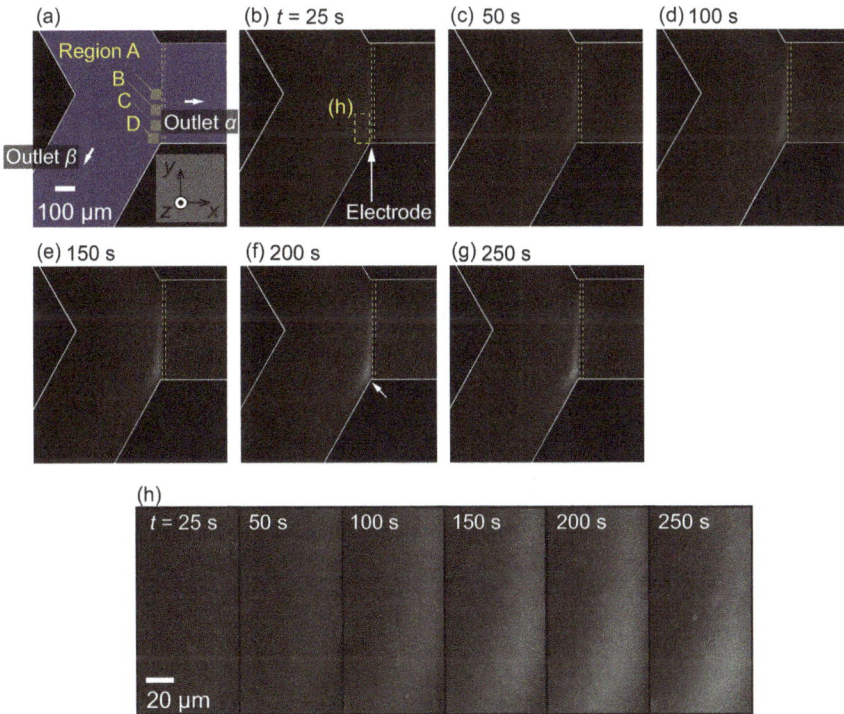

Figure 5. (**a**) Schematic figure of the test section and the positions of region A, B, C, and D analyzed in Figure 6; (**b–g**) Time series of the nanoparticle fluorescence. The particle flow separation is induced by microscale thermophoresis for the case with a particle diameter $d = 100$ nm and $\Delta P = 1.0$ Pa. At $t = 0$ s, the heating by the electrode is initiated. Particle flow from the inlet is separated at the Y-shaped branch. Because the thermophoresis is directed to the colder region, the PS particles cannot enter the outlet α. (**h**) Magnified figures of (**b–g**) for a rectangular region indicated in (**b**).

Comparing the results from microparticles, as shown in Figures 3 and 4, with those from nanoparticles, as shown in Figures 5 and 6, it can be concluded that the nanoparticles are more difficult to separate. This is because the thermal fluctuations become comparable to or stronger than the thermophoretic force and flow drag. If one wants to separate the nanoparticles more effectively, there are some methods of improvement. It should be noted that producing a larger temperature increase is not an appropriate option since it will easily cause the solution to boil if an aqueous solution is used. Firstly, downsizing the electrode width to produce a steeper temperature gradient will be effective. For this approach, the fabrication process of the electrode pattern should be improved to achieve a better yield ratio. Another approach is to use a different choice of electrolyte. It was reported in [34] that the Debye length, which is determined by the concentration of the electrolyte, affects the strength of the thermophoresis. Since the theory of thermophoresis has not been fully developed yet, such an approach will depend on trial and error. The present study can be used as a reference toward achieving more efficient nanoparticle separation using thermophoresis.

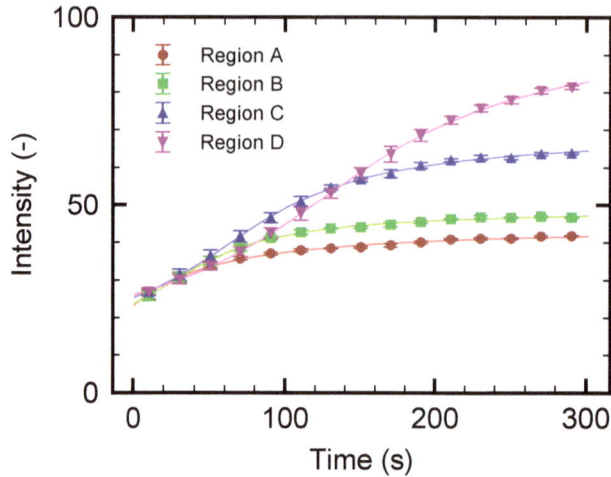

Figure 6. Time-development for fluorescence intensity of nanoparticles in the regions A, B, C, and D indicated in Figure 5a. The pressure difference ΔP is set to 1.0 Pa.

Next, a similar experiment for the nanoparticles using a smaller pressure difference $\Delta P = 0.5$ Pa is carried out with a longer experimental duration of 900 s. The intention here is to increase the nanoparticle concentration near the electrode using a slower mean flow of the continuous phase. It should be noted that the slower flow speed is expected to prevent the transport of the concentrated particles into the outlet β. The result is presented in Figure 7. The fluorescence in the left side of the electrode starts to increase as in the previous case with $\Delta P = 1.0$ Pa, as shown in Figure 5. At a first glance, the increase of the fluorescence intensity is enhanced due to the slower flow speed. However, it seems that the intensity increase converges to a constant value at a later stage of the experiment, as shown in Figure 7f–h. Then, the time-development of the fluorescence intensity in regions A, B, C, and D as indicated in Figure 7a is investigated. Figure 8 presents the intensity for these four regions. First, it is found that the intensity for region C is the most prominent. This trend is different from that in Figure 6, where the intensity of region D is the largest. This is attributed to the slower flow velocity component in the y direction, which transports the particles from region A to the negative y direction. The intensity for regions A, B, and D becomes saturated and fluctuated slightly. The fluctuation is caused by the occasional leak of trapped nanoparticles into the outlet α. Such a leak may be attributed to the longer experimental duration, which may cause an increase in the overall temperature in the entire microfluidic channel. Recall that the thermophoresis is driven by the temperature gradient, which may be diminished by the increase of the entire temperature. The result from region C, at which the increase of the intensity is the most prominent, shows that the intensity starts to decrease after $t = 300$ s. These results indicate that the effect of thermophoresis diminishes as time goes on, which also supports the speculation stated above regarding the entire temperature increase. To avoid the entire increase in the temperature, the use of different material for the channel wall is proposed. For instance, a Si substrate instead of the present glass substrate will dissipate the heat more efficiently, resulting in a more stable temperature field. A better design of the device which reduces such a leak will be explored in the future studies. Despite the remaining challenges, it is concluded that the concept of nanoparticle flow separation is demonstrated successfully using the present microfluidic device.

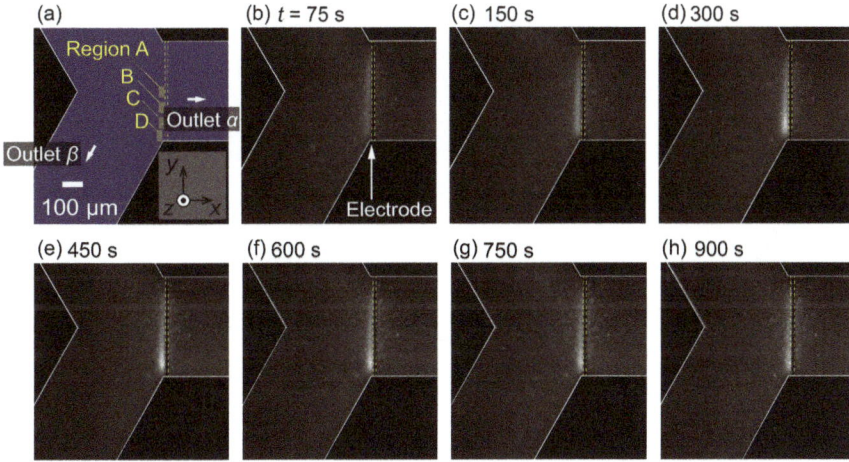

Figure 7. (a) Schematic figure of the test section and the positions of region A, B, C, and D analyzed in Figure 8; (b–h) Time series of the nanoparticle fluorescence. The particle flow separation is induced by microscale thermophoresis for the case with a particle diameter $d = 100$ nm and $\Delta P = 0.5$ Pa. At $t = 0$ s, the heating by electrode is initiated. Particle flow from the inlet is separated at the Y-shaped branch. Because the thermophoresis is directed to the colder region, the PS particles hardly enter the outlet α.

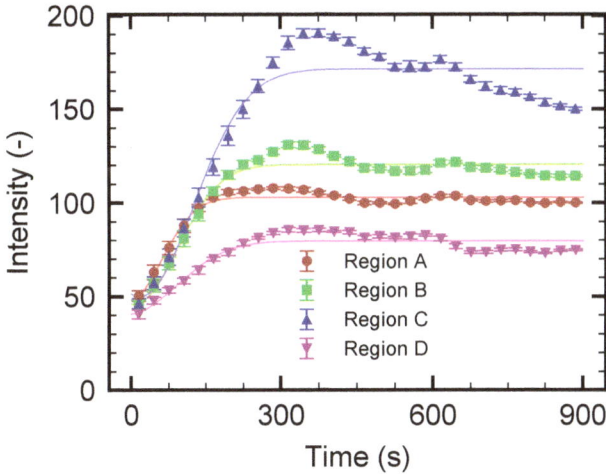

Figure 8. Time-development for fluorescence intensity of nanoparticles in regions A, B, C, and D indicated in Figure 7a. The pressure difference ΔP is set to 0.5 Pa.

3.4. Numerical Modeling for Nanoparticle Distribution

In this section, a simple model is proposed to explain the dynamics of the nanoparticle distribution observed in the experiment detailed in Figures 5 and 6. A rectangular region near the left side of the electrode is focused, as shown in Figure 9a. The width δ is smaller than w, and we consider the number density of the nanoparticles is uniform in the x direction. More precisely, a one-dimensional convection diffusion equation along the Y axis, that is, the left side of the electrode heater, is considered. It should be noted that Y is used for the 1D model in Figure 9a, instead of y in the original coordinates. Let $N(Y,t)$ be the number density of the nanoparticles with a time variable t. Velocity vectors of the continuous phase are denoted by (u_x, u_y) in the xy plane. These quantities are obtained from the experimental results and thus are given functions in the following numerical model. Due to the temperature gradient produced by the heater, thermophoresis of the particles with the velocity $u_T = -D_T(\frac{\partial T}{\partial x})$ is induced in the x direction, where D_T is the thermophoretic mobility. As described in the previous study [54], D_T was estimated to be almost 1×10^{-12} m$^2 \cdot$ s$^{-1} \cdot$K^{-1}, that is, the thermophoretic velocity was directed toward the colder region. The temperature gradient $\frac{\partial T}{\partial x}$ was estimated from the previous study [53] to be $\frac{\partial T}{\partial x} = 0.6 \times 10^6$ K$\cdot$m^{-1}. It should be noted that u_T is considered to be uniform with respect to Y, and the effective speed of particle transport in the x direction can be written as $U_x(Y) = u_x(Y) - u_T$. Using these notations, the governing equation for N can be written as

$$\frac{\partial N}{\partial t} + \frac{\partial(Nu_y)}{\partial Y} + D\frac{\partial^2 N}{\partial Y^2} = S(Y), \quad S(Y) = N_0 U_x(Y)/L_0, \tag{1}$$

where S is a source term that represents the supply of nanoparticles to the computational domain, N_0 is a uniform initial value of N, D is a diffusion coefficient, and L_0 is the length of the simulation region in the x direction. The flow velocity (u_x, u_y) is obtained from the PIV data. More specifically, the values presented in Figure 2a at $x = -100$ μm and $|y| < w/2 = 225$ μm are used, and a least-squares fit is made to determine (u_x, u_y). D is obtained from the Stokes–Einstein relation, that is, $D = k_B T/(3\pi \eta d)$, where $T = 360$ K is the temperature near the electrode heater at the steady state, $\eta = 3.2 \times 10^{-4}$ Pa$\cdot$s [60] is the viscosity at temperature T, and the particle diameter is $d = 100$ nm, as in the experiments. It should be noted that the steady temperature field is used for the modeling because an almost steady state was established after several seconds of heating [53].

Boundary conditions for Equation (1) are given as

$$\partial N/\partial Y = 0, \quad (Y = w/2), \tag{2}$$
$$\partial N/\partial t = S(Y), \quad (Y = w/2) \quad \rightarrow \quad N(Y = w/2) = N_0 + (N_0 U_x(Y)/L_0)t, \tag{3}$$

where Equation (3) is introduced by assuming that $\partial u_y/\partial Y|_{Y=w/2} = 0$. The boundary conditions (2) and (3) physically mean that N and (u_x, u_y) for $Y > w/2$, i.e., in the inlet channel, are uniform, respectively. Equation (3) results in the increase of particle density at a constant rate due to the source term.

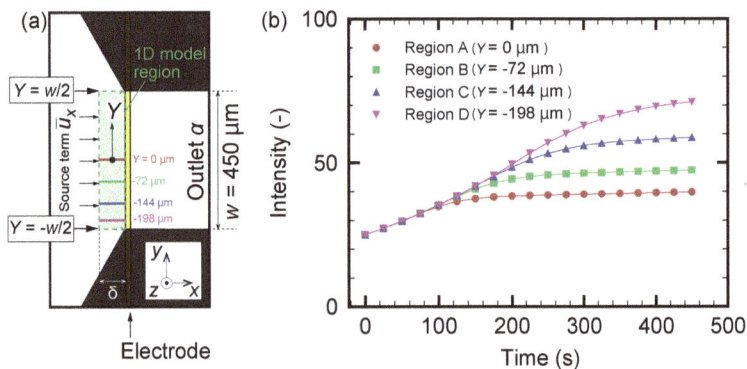

Figure 9. (a) Schematic of the numerical model on the concentration increase at the branch. (b) Numerical results regarding the time-development for fluorescence intensity of nanoparticles in the regions A ($Y \approx 0$ μm), B ($Y \approx -72$ μm), C ($Y \approx -144$ μm), and D ($Y \approx -198$ μm), shown in panel (a) and corresponding to Figure 6, where $Y = 0$ μm is placed at the center of the outlet α in the Y direction.

The initial condition is uniform with the bulk value $N(Y,0) = N_0$. Since the number of particles is linearly correlated to the fluorescence intensity in the experiments, N and N_0 can be interpreted as the fluorescence intensity at Y and t and that for the initial state. For L_0, the value $L_0 = 1.0 \times 10^4$ μm is used, which is chosen so that the model gives a result with a similar order of magnitude as the experiment. This is larger than the actual value of $O(100)$ μm, as shown in Figure 9a. The overestimation in the model may be due to the crude approximation made for the source term S. Nonetheless, the model is considered to be suitable for use as the first step in the qualitative investigation of the particle distribution made below. This model can be solved using a standard finite-difference scheme.

The numerical results are shown in Figure 9b, where the time-development of the intensity, which is scaled so that the initial value is consistent with the experimental results given in Figure 6, is presented for several positions Y. Here, $Y = 0, -72, -144$, and -198 μm correspond to regions A, B, C, and D in the experiments, as indicated in Figure 9a. By comparing the experiment in Figure 6 and the simulation in Figure 9b, a qualitative agreement is found; however, the range of times is different in these figures. From the simulation, the dynamic behavior of the nanoparticle concentration can be explained as follows. The nanoparticles are supplied to the left-side region of the electrode heater by the flow u_x of the continuous phase. Due to the thermophoresis, the nanoparticles cannot go beyond the heated electrode, and are concentrated in its left-side region. This is the reason for the overall increase in the intensity that can be observed in Figure 9b. As Y decreases, u_y also decreases, as shown in the PIV result of Figure 2a. Therefore, an increased particle concentration is transported in the negative Y direction. This leads to the larger concentration increase rate for region D compared with that for A. The present model indicates that the formation of a more concentrated region near the electrode is related to the spatial distribution of (u_x, u_y), and the design of the flow field at the branch is important for the control of the nanoparticle distribution.

4. Concluding Remarks

In the present study, a microfluidic channel with a Y-shaped branch, at which an inlet flow was separated into two symmetric outlet flows, has been developed. A thin-film electrode heater was fabricated at the entrance of one of the outlets to induce a local temperature rise for microscale thermophoresis of dispersed particles. Since the particles were repelled from the hot region, only the solvent entered the outlet with the heater, and the micro- and nanoparticles were transported to the other outlet. A simple model for the nanoparticle distribution based on the convection diffusion equation, which includes the effect of thermophoresis, was introduced, and a qualitative agreement with the experimentally observed nanoparticle motion was obtained. In this way, the particle flow separation using the microscale thermophoresis was demonstrated at the branched channel.

It resulted that the flow and temperature profiles were important to understand the detail of the particle behavior at the branch, such as the formation of a highly concentrated region. This was not shown in the previous study [53], and is an important finding of the present experiment. Therefore, the design of the device should be improved for the systematic investigation toward optimal particle flow control. Some fundamental research on flow separation, including the effect of heat transfer, was also presented in [61,62], where a backward facing step was considered instead of branched channels. In this research, the formation of recirculation zone, flow separation, and reattachment was investigated. The effect of thermophoresis on such flow geometries will be another interesting direction of thermophoretic particle flow separation. The theoretical aspects of the underlying physics of microscale thermophoresis and the above-mentioned systematic experiments will be carried out in future studies.

Author Contributions: T.T., K.D., and S.K. conceived and designed the research; T.T. and Y.M. constructed the measurement system; T.T., Y.M., and R.K. performed the experiments and analyzed the data; T.T., K.D., and S.K. wrote the paper.

Funding: This research is supported by the Japan Society for the Promotion of Science (JSPS) KAKENHI Grant No. JP18H05242 for Scientific Research (S) and JSPS KAKENHI Grant No. 18K13687 for Young Scientists.

Micromachines **2019**, *10*, 321

Conflicts of Interest: The authors declare no conflict of interest.

References

1.   Choi, S.U.S. Enhancing thermal conductivity of fluids with nanoparticles. In *Developments and Applications of Non-Newtonian Flows*; Siginer, D.A., Wang, H.P., Eds.; FED–Volume 231/MD–Volume 66; The American Society of Mechanical Engineers: New York, NY, USA, 1995; pp. 99–105.
2.   Yu, W.; Xie, H. A review on nanofluids: Preparation, stability mechanisms, and applications. *J. Nanomater.* **2012**, *2012*, 435873. [CrossRef]
3.   Islam, M.R.; Shabani, B.; Rosengarten, G. Nanofluids to improve the performance of PEM fuel cell cooling systems: A theoretical approach. *Appl. Energy* **2016**, *178*, 660–671. [CrossRef]
4.   Xiao, B.; Wang, W.; Zhang, X.; Long, G.; Chen, H.; Cai, H.; Deng, L. A novel fractal model for relative permeability of gas diffusion layer in proton exchange membrane fuel cell with capillary pressure effect. *Fractals* **2019**, *27*, 1950012. [CrossRef]
5.   Liang, M.; Liu, Y.; Xiao, B.; Yang, S.; Wang, Z.; Han, H. An analytical model for the transverse permeability of gas diffusion layer with electrical double layer effects in proton exchange membrane fuel cells. *Int. J. Hydrog. Energy* **2018**, *43*, 17880–17888. [CrossRef]
6.   Hossain, R.; Mahmud, S.; Dutta, A.; Pop, I. Energy storage system based on nanoparticle-enhanced phase change material inside porous medium. *Int. J. Therm. Sci.* **2015**, *91*, 49–58. [CrossRef]
7.   Xiao, B.; Wang, W.; Zhang, X.; Long, G.; Fan, J.; Chen, H.; Deng, L. A novel fractal solution for permeability and Kozeny-Carman constant of fibrous porous media made up of solid particles and porous fibers. *Powder Technol.* **2019**, *349*, 92–98. [CrossRef]
8.   Liang, M.; Fu, C.; Xiao, B.; Luo, L.; Wang, Z. A fractal study for the effective electrolyte diffusion through charged porous media. *Int. J. Heat Mass Trans.* **2019**, *137*, 365–371. [CrossRef]
9.   Ehtesabi, H.; Ahadian, M.M.; Taghikhani, V.; Ghazanfari, M.H. Enhanced heavy oil recovery in sandstone cores using TiO_2 nanofluids. *Energy Fuels* **2013**, *28*, 423–430.
10. Long, G.; Xu, G. The effects of perforation erosion on practical hydraulic-fracturing applications. *SPE J.* **2017**, *22*, 645–659. [CrossRef]
11. Long, G.; Liu, S.; Xu, G.; Wong, S.-W.; Chen, H.; Xiao, B. A perforation-erosion model for hydraulic-fracturing applications. *SPE Prod. Oper.* **2018**, *33*, 770–783. [CrossRef]
12. Xiao, B.; Zhang, X.; Wang, W.; Long, G.; Chen, H.; Kang, H.; Ren, W. A fractal model for water flow through unsaturated porous rocks. *Fractals* **2018**, *26*, 1840015. [CrossRef]
13. Tsutsui, M.; Taniguchi, M.; Yokota, K.; Kawai, T. Identifying single nucleotides by tunnelling current. *Nat. Nanotechnol.* **2010**, *5*, 286. [CrossRef]
14. Sackmann, E.K.; Fulton, A.L.; Beebe, D.J. The present and future role of microfluidics in biomedical research. *Nature* **2014**, *507*, 181. [CrossRef] [PubMed]
15. Stone, H.A.; Stroock, A.D.; Ajdari, A. Engineering flows in small devices: Microfluidics toward a lab-on-a-chip. *Annu. Rev. Fluid Mech.* **2004**, *36*, 381–411. [CrossRef]
16. Huang, L.R.; Cox, E.C.; Austin, R.H.; Sturm, J.C. Continuous particle separation through deterministic lateral displacement. *Science* **2004**, *304*, 987–990. [CrossRef] [PubMed]
17. Di Carlo, D.; Irimia, D.; Tompkins, R.G.; Toner, M. Continuous inertial focusing, ordering, and separation of particles in microchannels. *Proc. Natl. Acad. Sci. USA* **2007**, *104*, 18892–18897. [CrossRef]
18. Shintaku, H.; Imamura, S.; Kawano, S. Microbubble formations in MEMS-fabricated rectangular channels: A high-speed observation. *Exp. Therm. Fluid Sci.* **2008**, *32*, 1132–1140. [CrossRef]
19. Kuntaegowdanahalli, S.S.; Bhagat, A.A.S.; Kumar, G.; Papautsky, I. Inertial microfluidics for continuous particle separation in spiral microchannels. *Lab Chip* **2009**, *9*, 2973–2980. [CrossRef]
20. Shi, J.; Huang, H.; Stratton, Z.; Huang, Y.; Huang, T.J. Continuous particle separation in a microfluidic channel via standing surface acoustic waves (SSAW). *Lab Chip* **2009**, *9*, 3354–3359. [CrossRef]
21. Qian, W.; Doi, K.; Kawano, S. Effects of polymer length and salt concentration on the transport of ssDNA in nanofluidic channels. *Biophys. J.* **2017**, *112*, 838–849. [CrossRef]
22. Keyser, U.F. Controlling molecular transport through nanopores. *J. R. Soc. Interface* **2011**, *8*, 1369. [CrossRef] [PubMed]

23. Uehara, S.; Shintaku, H.; Kawano, S. Electrokinetic flow dynamics of weakly aggregated λDNA confined in nanochannels. *J. Fluids Eng.* **2011**, *133*, 121203. [CrossRef]

24. Sanghavi, B.J.; Varhue, W.; Chávez, J.L.; Chou, C.-F.; Swami, N.S. Electrokinetic preconcentration and detection of neuropeptides at patterned graphene-modified electrodes in a nanochannel. *Anal. Chem.* **2014**, *86*, 4120–4125. [CrossRef] [PubMed]

25. Tanaka, S.; Tsutsui, M.; Theodore, H.; Yuhui, H.; Arima, A.; Tsuji, T.; Doi, K.; Kawano, S.; Taniguchi, M.; Kawai, T. Tailoring particle translocation via dielectrophoresis in pore channels. *Sci. Rep.* **2016**, *6*, 31670. [CrossRef]

26. Shin, S.; Ault, J.T.; Warren, P.B.; Stone, H.A. Accumulation of colloidal particles in flow junctions induced by fluid flow and diffusiophoresis. *Phys. Rev. X* **2017**, *7*, 041038. [CrossRef]

27. Prieve, D.C.; Malone, S.M.; Khair, A.S.; Stout, R.F.; Kanj, M.Y. Diffusiophoresis of charged colloidal particles in the limit of very high salinity. *Proc. Natl. Acad. Sci. USA* **2018**. [CrossRef] [PubMed]

28. Shin, S.; Warren, P.B.; Stone, H.A. Cleaning by surfactant gradients: Particulate removal from porous materials and the significance of rinsing in laundry detergency. *Phys. Rev. Appl.* **2018**, *9*, 034012. [CrossRef]

29. Ault, J.T.; Shin, S.; Stone, H.A. Diffusiophoresis in narrow channel flows. *J. Fluid Mech.* **2018**, *854*, 420–448. [CrossRef]

30. Seki, T.; Okuzono, T.; Toyotama, A.; Yamanaka, J. Mechanism of diffusiophoresis with chemical reaction on a colloidal particle. *Phys. Rev. E* **2019**, *99*, 012608. [CrossRef]

31. Piazza, R. Thermophoresis: Moving particles with thermal gradients. *Soft Matter* **2008**, *4*, 1740. [CrossRef]

32. Piazza, R.; Parola, A. Thermophoresis in colloidal suspensions. *J. Phys. Condens. Matter* **2008**, *20*, 153102. [CrossRef]

33. Würger, A. Thermal non-equilibrium transport in colloids. *Rep. Prog. Phys.* **2010**, *73*, 126601. [CrossRef]

34. Duhr, S.; Braun, D. Why molecules move along a temperature gradient. *Proc. Natl. Acad. Sci. USA* **2006**, *103*, 19678. [CrossRef]

35. Iacopini, S.; Rusconi, R.; Piazza, R. "The macromolecular tourist": Universal temperature dependence of thermal diffusion in aqueous colloidal suspensions. *Eur. Phys. J. E* **2006**, *19*, 59. [CrossRef]

36. Ning, H.; Buitenhuis, J.; Dhont, J.K.; Wiegand, S. Thermal diffusion behavior of hard-sphere suspensions. *J. Chem. Phys.* **2006**, *125*, 204911. [CrossRef]

37. Vigolo, D.; Rusconi, R.; Stone, H.A.; Piazza, R. Thermophoresis: Microfluidics characterization and separation. *Soft Matter* **2010**, *6*, 3489. [CrossRef]

38. Eslahian, K.A.; Majee, A.; Maskos, M.; Würger, A. Specific salt effects on thermophoresis of charged colloids. *Soft Matter* **2014**, *10*, 1931–1936. [CrossRef]

39. Tsuji, T.; Kozai, K.; Ishino, H.; Kawano, S. Direct observations of thermophoresis in microfluidic systems. *Micro Nano Lett.* **2017**, *12*, 520. [CrossRef]

40. Lin, L.; Wang, M.; Peng, X.; Lissek, E.N.; Mao, Z.; Scarabelli, L.; Adkins, E.; Coskun, S.; Unalan, H.E.; Korgel, B.A.; et al. Opto-thermoelectric nanotweezers. *Nat. Photonics* **2018**, *12*, 195. [CrossRef]

41. Jiang, H.-R.; Wada, H.; Yoshinaga, N.; Sano, M. Manipulation of colloids by a nonequilibrium depletion force in a temperature gradient. *Phys. Rev. Lett.* **2009**, *102*, 208301. [CrossRef]

42. Maeda, Y.T.; Tlusty, T.; Libchaber, A. Effects of long DNA folding and small RNA stem–loop in thermophoresis. *Proc. Natl. Acad. Sci. USA* **2012**, *109*, 17972. [CrossRef] [PubMed]

43. Wienken, C.J.; Baaske, P.; Rothbauer, U.; Braun, D.; Duhr, S. Protein-binding assays in biological liquids using microscale thermophoresis. *Nat. Commun.* **2010**, *1*, 100. [CrossRef]

44. Seidel, S.A.I.; Wienken, C.J.; Geissler, S.; Jerabek-Willemsen, M.; Duhr, S.; Reiter, A.; Trauner, D.; Braun, D.; Baaske, P. Label-free microscale thermophoresis discriminates sites and affinity of protein–ligand binding. *Angew. Chem. Int. Edit.* **2012**, *51*, 10656. [CrossRef]

45. Burelbach, J.; Zupkauskas, M.; Lamboll, R.; Lan, Y.; Eiser, E. Colloidal motion under the action of a thermophoretic force. *J. Chem. Phys.* **2017**, *147*, 094906. [CrossRef]

46. Burelbach, J.; Brückner, D.B.; Frenkel, D.; Eiser, E. Thermophoretic forces on a mesoscopic scale. *Soft Matter* **2018**, *14*, 7446–7454. [CrossRef]

47. Burelbach, J.; Frenkel, D.; Pagonabarraga, I.; Eiser, E. A unified description of colloidal thermophoresis. *Eur. Phys. J. E* **2018**, *41*, 7.

48. Tsuji, T.; Saita, S.; Kawano, S. Thermophoresis of a Brownian particle driven by inhomogeneous thermal fluctuation. *Physica A* **2018**, *493*, 467. [CrossRef]

49. Galliéro, G.; Volz, S. Thermodiffusion in model nanofluids by molecular dynamics simulations. *J. Chem. Phys.* **2008**, *128*, 064505. [CrossRef] [PubMed]

50. Lüsebrink, D.; Yang, M.; Ripoll, M. Thermophoresis of colloids by mesoscale simulations. *J. Phys. Condens. Matter* **2012**, *24*, 284132. [CrossRef] [PubMed]

51. Tsuji, T.; Iseki, H.; Hanasaki, I.; Kawano, S. Molecular dynamics study of force acting on a model nano particle immersed in fluid with temperature gradient: Effect of interaction potential. *AIP Conf. Proc.* **2016**, *1786*, 110003.

52. Tsuji, T.; Iseki, H.; Hanasaki, I.; Kawano, S. Negative thermophoresis of nanoparticles interacting with fluids through a purely-repulsive potential. *J. Phys. Condens. Matter* **2017**, *29*, 475101. [CrossRef] [PubMed]

53. Tsuji, T.; Saita, S.; Kawano, S. Dynamic pattern formation of microparticles in a uniform flow by an on-chip thermophoretic separation device. *Phys. Rev. Appl.* **2018**, *9*, 024035. [CrossRef]

54. Tsuji, T.; Sasai, Y.; Kawano, S. Thermophoresithermophoretic manipulation of micro- and nanoparticle flow through a sudden contraction in a microchannel with near-infrared laser irradiation. *Phys. Rev. Appl.* **2018**, *10*, 044005. [CrossRef]

55. Briggs, J.A.G.; Grünewald, K.; Glass, B.; Förster, F.; Kräusslich, H.-G.; Fuller, S.D. The mechanism of HIV-1 core assembly: Insights from three-dimensional reconstructions of authentic virions. *Structure* **2006**, *14*, 15. [CrossRef]

56. Bouvier, N.M.; Palese, P. The biology of influenza viruses. *Vaccine* **2008**, *26*, D49. [CrossRef]

57. Kawaguchi, C.; Noda, T.; Tsutsui, M.; Taniguchi, M.; Kawano, S.; Kawai, T. Electrical detection of single pollen allergen particles using electrode-embedded microchannels. *J. Phys. Condens. Matter* **2012**, *24*, 164202. [CrossRef]

58. Bruus, H. *Theoretical Microfluidics*; Oxford University Press: Oxford, UK, 2007.

59. Mäki, A.-J.; Hemmilä, S.; Hirvonen, J.; Girish, N.N.; Kreutzer, J.; Hyttinen, J.; Kallio, P. Modeling and experimental characterization of pressure drop in gravity-driven microfluidic systems. *J. Fluids Eng.* **2015**, *137*, 021105. [CrossRef]

60. Kestin, J.; Sokolov, M.; Wakeham, W.A. Viscosity of liquid water in the range $-8\,°C$ to $150\,°C$ *J. Chem. Ref. Data* **1978**, *7*, 941–948. [CrossRef]

61. Abu-Nada, E. Numerical prediction of entropy generation in separated flows. *Entropy* **2005**, *7*, 234–252. [CrossRef]

62. Pour, M.S.; Nassab, S.G. Numerical investigation of forced laminar convection flow of nanofluids over a backward facing step under bleeding condition. *J. Mech.* **2012**, *28*, N7–N12. [CrossRef]

micromachines

MDPI

Article

Numerical and Experimental Analyses of Three-Dimensional Unsteady Flow around a Micro-Pillar Subjected to Rotational Vibration

Kanji Kaneko [1,2], Takayuki Osawa [2], Yukinori Kametani [2], Takeshi Hayakawa [1], Yosuke Hasegawa [2,*] and Hiroaki Suzuki [1,*]

[1] Faculty of Science and Engineering, Chuo University, Tokyo 112-8551, Japan; kaneko@nano.mech.chuo-u.ac.jp (K.K.); hayaka-t@mech.chuo-u.ac.jp (T.H.)
[2] Institute of Industrial Science, The University of Tokyo, Tokyo 153-8505, Japan; ta-osawa@iis.u-tokyo.ac.jp (T.O.); yukkam@iis.u-tokyo.ac.jp (Y.K.)
*   Correspondence: ysk@iis.u-tokyo.ac.jp (Y.H.); suzuki@mech.chuo-u.ac.jp (H.S.)

Received: 6 November 2018; Accepted: 13 December 2018; Published: 17 December 2018

Abstract: The steady streaming (SS) phenomenon is gaining increased attention in the microfluidics community, because it can generate net mass flow from zero-mean vibration. We developed numerical simulation and experimental measurement tools to analyze this vibration-induced flow, which has been challenging due to its unsteady nature. The validity of these analysis methods is confirmed by comparing the three-dimensional (3D) flow field and the resulting particle trajectories induced around a cylindrical micro-pillar under circular vibration. In the numerical modeling, we directly solved the flow in the Lagrangian frame so that the substrate with a micro-pillar becomes stationary, and the results were converted to a stationary Eulerian frame to compare with the experimental results. The present approach enables us to avoid the introduction of a moving boundary or infinitesimal perturbation approximation. The flow field obtained by the micron-resolution particle image velocimetry (micro-PIV) measurement supported the three-dimensionality observed in the numerical results, which could be important for controlling the mass transport and manipulating particulate objects in microfluidic systems.

Keywords: vibration-induced flow; micro-pillar; numerical analysis; micro-PIV; acoustofluidics

1. Introduction

The hydrodynamic phenomenon known as steady streaming (SS) is gaining increased attention for controlling flows and associated transport and mixing of chemical species as well as micro objects such as functionalized particles and cells in microfluidic devices [1–12]. This term represents the time-averaged non-zero mean flow induced by relative periodic oscillation with zero mean between the substrate and the adjacent bulk fluid. When an infinite planar substrate is oscillating in parallel to its surface, only a transient velocity field with zero mean is generated within a thin layer whose length scale is characterized by the Stokes boundary layer thickness represented as $\delta_s \sim (2\nu/\omega)^{1/2}$, where ν and ω are the kinematic viscosity and angular frequency, respectively [13]. However, when an obstacle is present in the flow field, the interaction between the oscillating bulk fluid and the obstacle creates vorticity and causes net-momentum transfer. This is similar to the Reynolds shear stress that arises from the correlation of the fluctuating velocity components in turbulent flows, and it appears as an additional forcing term in the averaged Navier-Stokes equations. As a result, the steady time-averaged velocity is generated, although the applied periodic forcing does not have a mean component. Because it requires no net displacement or a pressure gradient to drive the flow, the SS is expected to simplify and miniaturize microfluidic systems without introducing external pumps or tubing.

Despite its simplicity, prediction of the flow field induced by SS is nontrivial. Classically the SS flow fields around simple objects, such as a sphere and a cylinder, induced by the oscillatory motion relative to a surrounding fluid were studied [14,15]. The analytical solutions can be obtained in these cases. However, for practical microfluidics applications, a situation could be more complicated; the presence of (often non-straight) channel walls and obstacles with complex shapes does not allow us to derive analytical solutions. To date, several research groups have tackled this problem using numerical analysis. The group of Dr. Schwartz carefully studied the uni-directional oscillating flow in a straight channel with the rectangular cross section, in which an array of cylindrical posts is placed [16,17]. For the numerical analysis, they employed a perturbation approach under two-dimensional flow assumption, in which the periodically oscillating and steady flows were solved separately. Their experimental and numerical results showed good agreement in terms of the streamlines. However, since their main interest was in predicting the location of the center of eddies, into which small particles were trapped, the detailed velocity profile was not fully examined. One of the co-authors of the present work (Dr. Hayakawa) showed that circular vibration, instead of unidirectional vibration, induced the circularly rotating mean flow around a cylindrical micro-pillar. He and his colleagues utilized this phenomenon for manipulation and trapping of cells [7–9]. They numerically calculated the flow field using the perturbation approach under the two-dimensional assumption, and obtained peak velocity values in the profile that matched the experimental observations. In addition, the group of Dr. Costanzo established a numerical model to predict the flow in the acoustic-driven micromixing device developed by Dr. Huang in the same university [18–20]. Since the frequency of acoustic excitation is higher than SS, they considered the compressibility of the fluid, but the model was based on the perturbation approach under the two-dimensional assumption.

Although numerical results in the abovementioned studies were able to reproduce the experimentally observed SS flow fields, they are based on two assumptions; (1) the amplitude of perturbation (*s*) is small compared to the characteristic length of the system (e.g., the radius of the cylinder *a*), and (2) the flow is two-dimensional. However, the practical operational conditions of SS device may not be limited to the assumed condition of $s/a << 1$. Furthermore, in most microfluidic devices, the length-scale perpendicular to the substrate (e.g., the height of channels and structures) is generally much smaller than that in horizontal directions (e.g., the width and length of channels). In such cases, the Stokes layers developing from both top and bottom boundaries are not negligible, so that two-dimensional flow assumption breaks down in most of the flow domain [21]. The three-dimensionality of the SS flow should also arise when the obstacles have 3D shapes. So far, the assessment of the numerical results has mainly been limited to qualitative comparison to the streamlines, which can be readily obtained by the long-time exposure images of the tracer fluorescent beads in experiments. However, the streamlines as well as the magnitude of velocity should strongly depend on the height. To fully predict the SS flows and optimize them, an analytical tool that can directly simulate the 3D field without assumptions is needed.

More recently, several groups reported the 3D numerical simulations of SS flows. Amit et al. calculated the flow around a moving boundary by a commercial solver based on the finite element method [22]. The comparison between the numerical results and the experimental particle image velocimetry (PIV) measurement around a vibrating long cantilever showed quantitative agreement of the velocity field. Rallabandi et al. analyzed the 3D flow field induced by the acoustic actuation of a microbubble, and verified the results through comparison with astigmatism particle tracking velocimetry (APTV) measurements, although the analysis was still based on the small perturbation assumption [23].

Based on the above background, we developed a versatile numerical tool to calculate the 3D SS flow without assuming small perturbation and two-dimensionality of the flow in order to examine the flow induced around a circularly vibrating cylinder placed in a quiescent fluid between two parallel substrates. We employed a Lagrangian approach, in which the coordinate system of the simulation is fixed to the moving (vibrating or oscillating) substrate, instead of an Eulerian approach, in which the boundary is moving relative to the stationary reference frame. The governing equations of an incompressible fluid, i.e., Navier-Stokes and continuity equations, in which a temporally periodic inertia force due to the circular vibration was included, were directly solved by a pseudo-spectral method [24]. Once the numerical results in the moving (Lagrangian) coordinate were obtained, the flow field was converted to the stationary (Eulerian) coordinate. This approach enabled us to calculate the SS flow field without imposing the moving boundary nor the small perturbation approximation. A fluid-solid boundary is expressed by the level-set function [25], which is defined as a signed distance function from the surface. This allows to immerse an arbitrary 3D shape in the Eulerian coordinate system and a no-slip condition at a solid surface is achieved by a volume penalization method (VPM) [26]. Such an approach has a great advantage in implementing arbitrary complex obstacles in the fluid domain without generating numerical grids for each geometry. After obtaining the periodically varying instantaneous velocity field, the time-averaged velocity field of Stokes drift was obtained by tracking the virtual fluid particle imposed within the fluid.

The 3D SS velocity field obtained by the numerical simulation was validated by quantitatively comparing with the 3D micro PIV measurement results obtained from the confocal microscopy equipped with a high-speed camera. Our results exhibited the good agreement even within the steeply varying velocity profile within the Stokes layer. The 3D paths of tracer particles induced by the vortex structure adjacent to the edge of the cylinder observed in the experiment were reproduced in the numerical simulation.

2. Numerical Procedures

2.1. Computational Domain and Governing Equations

In this study, we calculated the flow field around a cylindrical micro-pillar placed between two parallel plates (Figure 1a). Both plates and a pillar fixed to the bottom plate periodically oscillate following a circular path parallel to the substrate (Figure 1b). To simulate this system, we employed a moving coordinate system that moves with the plates instead of the stationary coordinate system with moving boundaries. The flow field obtained in the moving coordinate system was eventually converted to the coordinate system at rest. The advantage of this approach is that the calculation code is simple and numerical accuracy is high, since boundaries (the substrate and the pillar) are stationary with respect to the coordinate system. The effect of the circular vibration was considered by applying an inertia force that rotates in accordance with the acceleration/deceleration of the moving coordinate.

Assuming that the liquid is incompressible and a Newtonian fluid, the liquid flow is governed by the following Navier-Stokes and the continuity equations:

$$\frac{\partial \underline{u_i}^*}{\partial t^*} + \underline{u_j}^* \frac{\partial \underline{u_i}^*}{\partial \underline{x_j}^*} = -\frac{1}{\rho^*} \frac{\partial \underline{p}^*}{\partial \underline{x_i}^*} + \nu^* \frac{\partial^2 \underline{u_i}^*}{\partial \underline{x_j}^* \partial \underline{x_j}^*} \tag{1}$$

$$\frac{\partial \underline{u_i}^*}{\partial \underline{x_i}^*} = 0 \tag{2}$$

Here, variables with an asterisk represents a dimensional quantity. The fluid kinetic viscosity and density are denoted by ν and ρ, respectively. The two orthogonal directions tangential to the substrate are set as $\underline{x}$ and $\underline{z}$, while the wall-normal direction is $\underline{y}$. The origin is located at the center of the bottom substrate. Time is defined as $\underline{t}$. The under-bar indicates a physical quantity in the coordinate system at rest, which is introduced so as to be distinguished from the moving coordinate along a circular orbit defined later. The fluid velocity and static pressure are denoted by $\underline{u_i}$ and p, respectively, where the subscript of i represents the three directions, namely, $i = 1, 2, 3$ corresponds to $\underline{x}, \underline{y}, \underline{z}$, respectively. A micro-pillar was attached to the bottom substrate and its height was set to be equal to the channel half height δ^*. No-slip conditions were applied at the bottom and top walls, i.e., $y^* = 0$ and $2\delta^*$, as well as the surface of the pillar. Periodic boundary conditions were employed in the x and z directions.

This condition corresponds to the case where the geometry shown in Figure 1a repeats in these two directions. The present configuration is chosen because it is relatively simple (the two parallel walls and the periodic placement of cylinders between the walls), while the truncation of the cylinder at the middle of the channel causes complex three-dimensionality of the resulting flow.

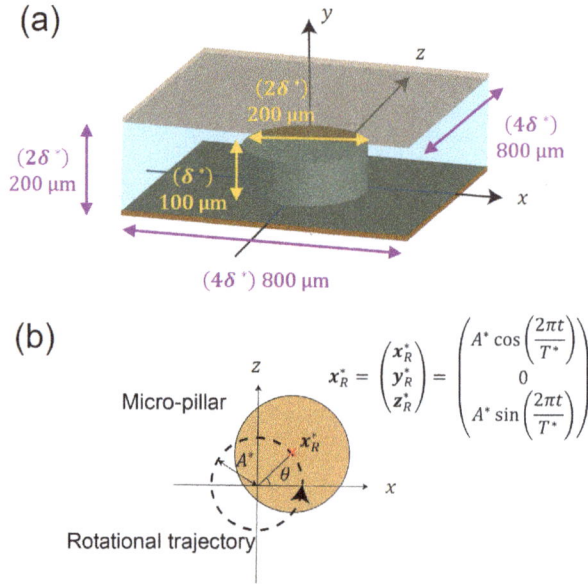

(a)

(b)

$$x_R^* = \begin{pmatrix} x_R^* \\ y_R^* \\ z_R^* \end{pmatrix} = \begin{pmatrix} A^* \cos\left(\frac{2\pi t}{T^*}\right) \\ 0 \\ A^* \sin\left(\frac{2\pi t}{T^*}\right) \end{pmatrix}$$

Figure 1. (**a**) Schematic of coordinate system and computational domain. (**b**) Circular vibration of the micro-pillar.

In the present study, the solid substrate with micro-pillars oscillates along a given circular tangential orbit as shown in Figure 1b. When the radius and period of the circular oscillation are expressed by A^* and T^*, the relative displacement of the substrate x^*_R is given as

$$x_R^* = \begin{pmatrix} x_R^* \\ y_R^* \\ z_R^* \end{pmatrix} = \begin{pmatrix} A^* \cos\left(\frac{2\pi t}{T^*}\right) \\ 0 \\ A^* \sin\left(\frac{2\pi t}{T^*}\right) \end{pmatrix} \tag{3}$$

Accordingly, the velocity of the solid substrate, u^*_R, can be obtained from the time derivative of x^*_R as

$$u_R^* = \begin{pmatrix} \frac{dx_R^*}{dt} \\ \frac{dy_R^*}{dt} \\ \frac{dz_R^*}{dt} \end{pmatrix} = \begin{pmatrix} -\frac{2\pi A^*}{T^*} \cos\left(\frac{2\pi t}{T^*}\right) \\ 0 \\ \frac{2\pi A^*}{T^*} \sin\left(\frac{2\pi t}{T^*}\right) \end{pmatrix} \tag{4}$$

Solving Equations (1) and (2) under the boundary condition of Equation (4) on the surfaces of the top and bottom substrates and the pillar requires the treatment of moving boundaries. In order to avoid it, a new coordinate system $x^* = (x^*, y^*, z^*)^T$, which moves with the same speed as the substrate, is introduced. With the reference length scale of δ^* and the oscillation period of T^*, the generalized dimensionless forms of Equations (1) and (2) on a rotating frame are expressed as

$$\frac{\partial u_i}{\partial t} + u_j \frac{\partial u_i}{\partial x_j} = -\frac{\partial p}{\partial x_i} + \frac{1}{Re} \frac{\partial^2 u_i}{\partial x_j \partial x_j} + f_i \tag{5}$$

$$\frac{\partial u_i}{\partial x_i} = 0 \tag{6}$$

where the fluid velocity relative to the substrate motion is defined as $u_i = \underline{u}_i - u_{R_i}$. The effect of the rotational vibration appears as a dimensionless inertia force f_i due to the acceleration/deceleration of the moving coordinate, which is given by

$$f_i = \frac{2\pi}{St} \begin{pmatrix} \cos(2\pi t) \\ 0 \\ \sin(2\pi t) \end{pmatrix} \tag{7}$$

The detailed derivation of Equations (5)–(7) can be found in the Supplementary Materials.

Here, the Strouhal number St is defined as the ratio of the time scales of the flow and the oscillation. Considering that the time scale of flow is given by δ^*/U^*_{max}, where $U^*_{max} = 2\pi A^*/T^*$ is the maximum velocity of the substrate, the Strouhal number is given by

$$St = \frac{\frac{\delta^*}{U^*_{max}}}{T^*} = \frac{\delta^*}{2\pi A^*} \tag{8}$$

The Reynolds number is given by

$$Re = \frac{\delta^{*2}}{\nu^* T^*} \tag{9}$$

The above two dimensionless parameters characterize the flow field considered in the present study. Since we solved the velocity field in the reference frame moved with the substrate, the top and bottom walls as well as the micro-pillar stay at rest, and therefore all the velocity components on these boundaries became null. In order to impose a no-slip condition at a solid surface with arbitrary geometry, we used a volume penalization method introduced in the next subsection.

2.2. Volume Penalization Method

One of the main objectives in the present study is to develop a numerical code that is capable of simulating a flow around pillars with arbitrary shapes. A volume penalization method is a kind of immersed boundary techniques, in which a complex structure is embedded in the Cartesian coordinate system. In contrast to using a boundary-fitted coordinate system, the immersed boundary technique has the advantages that grid generation is quite straightforward regardless of the complexity of the geometry and highly accurate discretization schemes developed for the Cartesian grid system can be applicable.

In the present study, the geometry of the pillar was first expressed by a level-set function in the Cartesian coordinate system. The level-set function ϕ_0 is a signed distance function from a surface [25] and it has been widely used to represent complex interface geometry. Here, ϕ_0 was defined positive inside the solid, and negative in the fluid domain. Then, we converted ϕ_0 to the phase-identification function ϕ, which was $\phi = 0$ inside the fluid, whereas $\phi = 1$ in the solid. In order to avoid numerical instability, the phase-identification function smoothly changes from zero to one across the interface within a few grid points. Specifically, the level-set function is converted to the phase-identification function by the following formulas:

$$\phi = 0 \qquad \phi_0 < -\delta_{int} \tag{10}$$

$$\phi = \left[1 + \exp\left\{\frac{4(\phi_0/\delta_{int})}{(\phi_0/\delta_{int})^2 - 1}\right\}\right]^{-1} \qquad -\delta_{int} < \phi_0 < \delta_{int} \tag{11}$$

$$\phi = 1 \qquad \delta_{int} < \phi_0 \tag{12}$$

The above function was chosen because it is differentiable within the entire domain, while it is exactly zero and unity in the fluid and solid domain, respectively.

In the volume penalization method, a no-slip condition at the solid surface is realized by introducing an artificial damping force to the Navier-Stokes equation (Equation (5)) as follows:

$$\frac{\partial u_i}{\partial t} + u_j \frac{\partial u_i}{\partial x_j} = -\frac{\partial p}{\partial x_i} + \frac{1}{Re} \frac{\partial^2 u_i}{\partial x_j \partial x_j} + f_i - \eta \phi u_i \tag{13}$$

Here, the final term on the right-hand side is the volume penalization term. Obviously, this term has a non-zero value and acts to suppress all the velocity components inside the solid, while Equation (13) reduces to the original Navier-Stokes equation (Equation (5)) within the flow domain where $\phi = 0$.

The advantage of the volume penalization method is that solid objects with different shapes can be easily implemented by changing the spatial distribution of ϕ in the same Cartesian grid system. The drawback is that the grid convergence is relatively slow, since the interface is not explicitly captured and smeared within a few grid points as mentioned above. In the present study, we made grid convergence tests and confirmed that the present results do not change significantly by refining the mesh further (Figure S1).

2.3. Numerical Methods and Conditions

We solved Equations (6) and (13) to obtain the velocity field in the moving coordinate by a pseud-spectral method, in which a solution was expanded by Fourier modes in the x, z direction and Chebyshev polynomials in the y direction, respectively. For time advancement, the Crank-Nicolson was used for diffusion terms, whereas the second-order Adams Bashforth scheme is applied for the convection terms. The inertia and VPM terms appearing in the third and fourth terms are taken into account with the Euler explicit method. The preset code was validated and successfully applied to control and estimation of unsteady turbulent flows in previous studies [24,27].

In this study, we set the diameter and height of micro-pillar to be 200 and 100 μm, respectively. The diameter of the pillar was shown to have only minor effect on the radial profile of the induced flow using the perturbation theory [8], so we studied the flow around the pillar with this representative diameter. The width of the computational domain was $4\delta^* = 800$ μm, which is equal to the spacing of pillars in the experiment. This dimension was set to be large enough so that induced flow profiles of neighboring pillars do not interact. The height of the domain was $2\delta^* = 200$ μm. The numbers of modes employed in the current simulation were $(N_1, N_2, N_3) = (64 \times 33 \times 64)$ in x, y, z directions, respectively. 3/2 rule was used for removing aliasing errors, so that the non-linear terms were evaluated in 1.5 times finer physical grid points in each direction. Throughout this work, the vibration amplitude was $A = 4$ μm and frequency was $f = 1000$ Hz. Accordingly, dimensionless numbers were $Re = 10$ and $St = 12.4/\pi$, respectively. The numerical time step was set to be $\Delta t = 1.0 \times 10^{-4}$, which indicates that $t^{-1} = 10^4$ time steps are required to compute the velocity field for one oscillation period ($t = 1$). The computation was started from a stationary flow at $t = 0$, and the rotational vibration was applied for $t = 60$ to achieve a fully developed velocity field. After the transient period, the flow field became completely periodic in one oscillation cycle. All statistics shown below were obtained by integrating the velocity data over one oscillation period after the flow field had reached the statistically steady state.

2.4. Derivation of Steady Streaming Flow (Time-Averaged Velocity Field)

It is widely known that the SS flow field could be essentially different depending on whether the averaging is made at a fixed stationary location (Eulerian frame) or along a particle moving with the local fluid velocity (Lagrangian frame) due to its oscillatory nature [20,23]. Therefore, we examined and compared the time-averaged SS velocity fields obtained in the Eulerian and Langrangian frames. In the Eulerian approach, the average velocity field was calculated by simply averaging the vector at identical positions in the stationary coordinate system converted from the moving coordinate system.

In the Lagrangian approach, trajectories of virtual fluid particles initially located at a uniform spacing of $\Delta = 8.33$ μm in the 3D flow field were calculated using the 4th order Runge-Kutta method [28]. The instantaneous velocities between the grids were linearly interpolated from the velocity of the surrounding 6 grid points. The instantaneous velocity field data during one circulating period consisted of 50 time frames, and the velocities at the time points between frames were also linearly interpolated using two neighboring frames. After tracking for five periods of the rotational vibration, the velocity field was obtained from the displacement vectors that connect the start and end points at the identical phase.

3. Experimental Procedure

3.1. Fabrication of Micro-Pillar Array

We fabricate the 5×5 array of cylindrical pillars with 200 μm diameter and 100 μm height, with 800 μm center-to-center intervals in accordance with the numerical simulation described in Section 2, using poly-dimethylsiloxane (PDMS) as the material (Figure 2). Four cylindrical pillars with 1.4 mm diameter and 200 μm height were arranged at the four corners of the substrate as spacers to determine the height of the fluid volume. In practice, the master mold was fabricated on a 2-inch silicon wafer by the deep reactive ion etching apparatus (RIE-400iPB, Samco, Japan) using the Bosch process at a rate of 0.4 μm/cycle. PDMS resin (KE-106, Shin-Etsu Chemical, Japan) mixed with its curing agent at 10:1 weight ratio was poured into the master mold. After curing at 50 °C for 120 min, the PDMS substrate was obtained by peeling it off from the mold (Figure 2b).

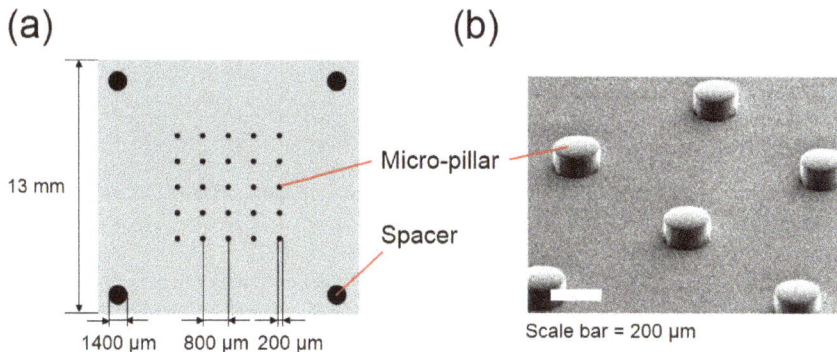

Figure 2. (**a**) Arrangement of the micro-pillars on a substrate. (**b**) SEM image of the micro-pillars.

3.2. Experimental Setup and Conditions

Since the PDMS is hydrophobic, air bubbles are often trapped around pillars when the liquid (water) is dropped on its surface. Thus, the substrate was made hydrophilic by the oxygen plasma (SEDE-GE, Meiwafosis Co., Ltd., Japan) treatment for 5 min. Immediately after this treatment, 10 μL of deionized (DI) water containing 0.5 μm yellow-green fluorescent beads (F8813, Thermo Fisher Scientific Inc., MA, USA) as a tracer was dropped to the center of the substrate. Then, the substrate was covered with a cover glass. The thickness of the fluid layer was set to 200 μm by spacers (Figure 3a). Then this assembly was fixed to the XY piezo-drive stage (ML-20XYL, MESS-TEK, Japan) (Figure 3b) using small slips of double-sided tape.

(a)

Cover glass

Droplet

Pillar plate

Sideview of assembly

(b)

Pillar plate assembly   Piezo-electric stage

Objective lens

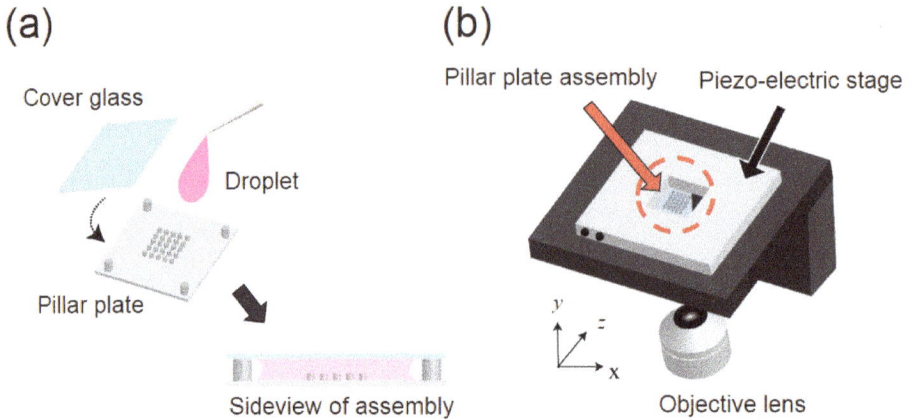

Figure 3. (**a**) Assembly of the micro-pillar plate. (**b**) Experimental setup for application of circular vibration.

To generate the circular vibration, sinusoidal wave signals with 90° phase offset was applied to the piezo stage using the waveform generator (AG 1022F, OWON, China) via the amplifier (M 2501-1, MESS-TEK, Japan). Applied voltage at 60 V induced $A = 4$ μm displacement over a wide range of frequency below the resonance of this actuator, which was confirmed from the image of the high-speed camera (Mi-2000, Photron, Japan). The vibration frequency was set at $f = 1000$ Hz throughout the present study in accordance with the numerical simulation.

3.3. PIV System

We used the confocal micro-PIV technique to measure the flow field around a micro-pillar. Figure 4 shows the schematic diagram of the confocal micro-PIV system. In this system, sequential images of fluorescence tracer particles are obtained by a high-speed camera (Mi-2000, Photron, Japan) via the high NA objective water immersion lens (XLUMPLFLN 20 XW, OLYMPUS, Japan) and a confocal scanner (CSU-X1, YOKOGAWA, Japan), and are stored in the PC. The continuous wave (CW) blue laser (488 nm, Sapphire SF, COHERENT, CA, USA) was used as the illumination light source. Since the frequency of the micro-pillar was 1000 Hz, the shutter speed was set to 1/21,000 s (~48 μs) so that the instantaneous particle images could be resolved without blurring. The frame rate of the image acquisition was set to 2000 fps.

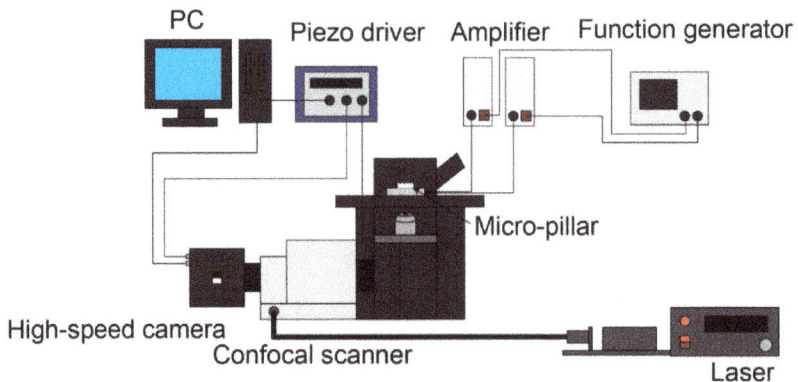

PC   Piezo driver   Amplifier   Function generator

Micro-pillar

High-speed camera
Confocal scanner

Laser

Figure 4. Schematic diagram of the confocal micro-PIV system.

3.4. PIV Analysis

Based on the acquired images, we obtained a two-dimensional velocity field using PIV analysis software (Koncerto II, Seika Digital Image, Japan). For obtaining the velocity vector, a recursive cross-correlation method was used, with a 8 × 8 pixel interrogation window and 50% overlap. This window size corresponds to 6.4 μm × 6.4 μm in the physical dimension. In the post processing, standard deviation validation and median filter were used to remove incorrect vectors. Since the micro-pillar oscillates at 1000 Hz, two images were recorded during the one cycle of a pillar rotation when the frame rate was 2000 fps. The SS velocity field was obtained from the displacement of tracer images at every two frames; i.e., images at identical rotational phase. Finally, the average velocity field induced around the pillar was obtained by averaging the flow fields of the 60 rotational cycles. We confirmed that the resultant average profile was almost independent of the interrogation window size (Figure S2).

3.5. Horizontal Visualization

A horizontal view of the trajectory of tracer particle was obtained through the objective lens with a long working distance (PAL-10-A, x10, SIGMAKOKI CO. LTD., Japan; WD = 34 mm) located on the side of the pillar substrate and captured by a monochrome CCD camera (BFS-U3-32S4M-C, FLIR Systems, Inc., OR, USA). Polystyrene beads with 10 μm diameter (01-00-104, micromod Partikeltechnologie GmbH., Germany) were used as tracer particles for easy visualization with low magnification lens. The sedimentation velocity of 10 μm bead with a specific density of 1.03 in water is estimated to be around 6.5 μm/s, so that the sedimentation distance within 1.7 s (duration of observation) is not significant.

4. Numerical Results

4.1. Instantaneous Velocity Field

In the present numerical simulation, a 3D instantaneous flow field around the pillar was obtained after converting the results in the moving coordinate to the stationary one. Figure 5 shows a two-dimensional (2D) vector plot in the plane 0.5δ away from the non-slip boundary at the bottom ($y = 50$ μm; the center plane of the pillar in y direction). The vectors and color maps in Figure 5 represent the flow direction and the absolute values of the horizontal component of the velocity ($|V_{hor}| = \sqrt{u_1^2 + u_3^2}$), respectively. The region near the pillar is enlarged while the calculation area is 800 μm for x and z directions. Figure 5a–d show the instantaneous fields with the phase shift of every 90° (see Movie S1 for all velocity fields within one rotation). Each flow field is rotationally symmetric, showing that the flow field reaches a fully developed state. High-speed regions appeared at the front and back faces of the cylindrical pillar in the traveling direction. These high-speed regions separate near the two sides of the pillar and cause a pair of vortex-like structures. The maximum fluid velocity in this region is ~ 25 mm/s, which is comparable to the velocity of oscillation $U_{max} = 2\pi Af$ ~ 25.2 mm/s. Although not shown in the figure, the comparable speed was also observed at the vicinity of no-slip boundaries (e.g., the upper and lower walls).

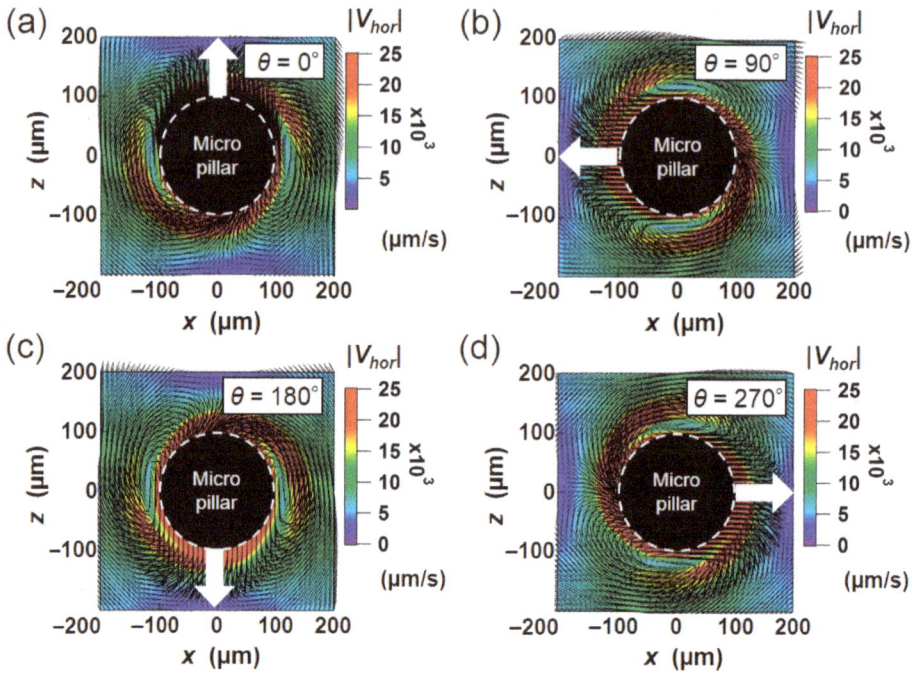

Figure 5. 2D vector plot of the instantaneous velocity field at phases of (**a**) 0°, (**b**) 90°, (**c**) 180°, (**d**) 270° within the one rotational cycle. White arrows indicate the instantaneous moving directions of the pillar.

4.2. Time-Averaged Flow Field

In a periodically vibrating flow field, the time-averaged velocity becomes null if there is no obstacle in the oscillating direction. However, a non-zero net velocity field appears when the obstacle exists as a result of the Stokes drift. As mentioned previously, we examined the two time-averaging approaches, i.e., Eulerian and Lagrangian averaging. In the latter, we calculated the velocity vectors from the displacement of the virtual tracer particles. An example of the 2D trajectory of a tracer, initially placed at $(x, y, z) = (150, 50, 0 \ \mu m)$ (horizontally 50 μm away from the side wall of the pillar) is shown in Figure 6. The tracer particle moves along the distorted orbital path, which can be seen as the superposition of the circular periodic motion and the steady translational movement toward the upper left in the figure. When the pillar is absent, the trajectory draws a perfect circle and returns to the original position after one rotation. Due to the existence of the pillar, however, the position in the same phase is shifted after each rotation. This displacement divided by the time of one rotation period corresponds to the mean velocity observed in experimental particle tracking generated in the SS flow field.

The two-dimensional vector plots of the averaged velocity field obtained by Eulerian and Lagrangian methods at the $y = 50 \ \mu m$ horizontal plane are depicted in Figure 7. The color map represents the absolute value of the horizontal component $| V_{hor} |$ of the average velocity. It is clear that a net flow is induced around the pillar in both cases. However, the radial peak position of the velocity is closer to the pillar in the former case ($r \sim 120 \ \mu m$ in the Eulerian method, but $r \sim 130 \ \mu m$ in the Lagrangian method). Moreover, the peak value is about two times greater in Eulerian method compared to the Lagrangian method. The result clearly indicates that the averaged flow field depends on the averaging methods. Because the mass transport and the paths of suspended molecules/particles are governed by the Lagrangian trajectories, the Lagrangian averaging is necessary to predict the above-mentioned transport phenomena in micro devices. In the following sections, the Lagrangian velocity obtained from the present simulation will be compared with the averaged translational velocity of fluorescence particles in the experiment in order to validate the present numerical code.

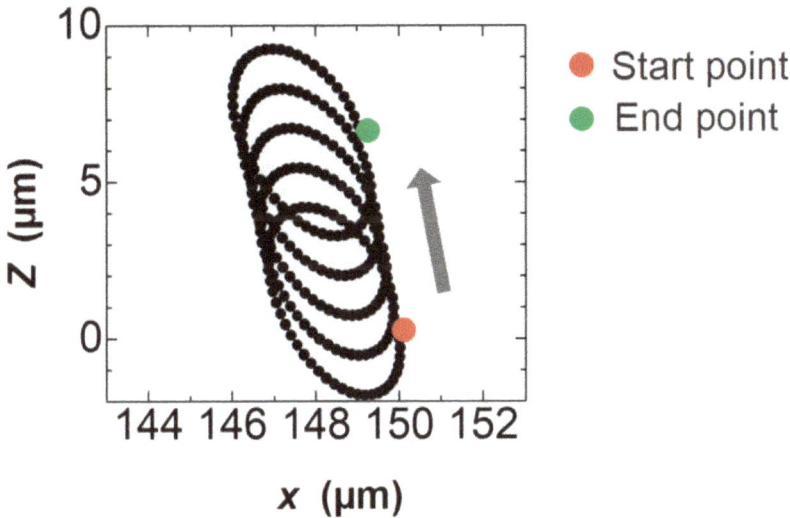

Figure 6. A 2D trajectory of a particle during five rotations.

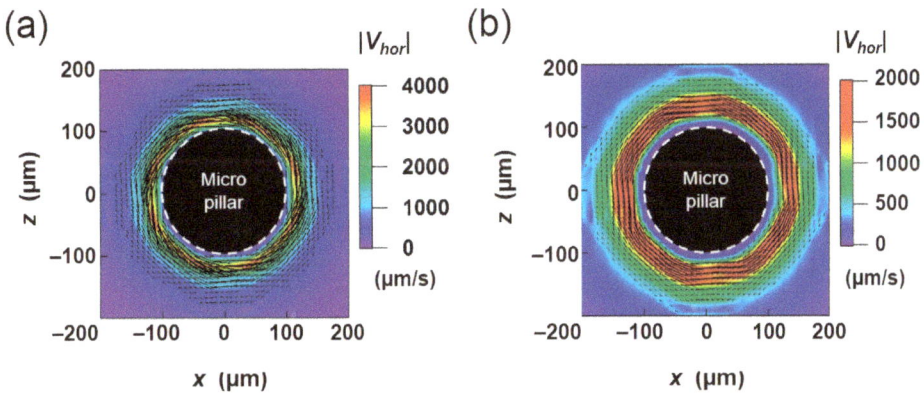

Figure 7. 2D vector plots of the averaged velocity flow field calculated by the (**a**) Eulerian and (**b**) Lagrangian approaches.

5. Comparison with experimental results

5.1. Results of PIV Measurement

The image of the movement of fluorescent tracer particles obtained at the $y = 50$ μm horizontal plane is shown in Figure 8a. Sixty instantaneous images at an identical rotational phase, captured by a high-speed camera, are superimposed. Although tracer beads actually moved along the rotating trajectory in accordance with the rotational vibration, the net displacement along the pillar sidewall can be clearly seen by connecting the positions at the same phase. The result of PIV analysis, obtained in this series of images, is depicted in Figure 8b. The condition is the same as that in the numerical result shown in Figure 7. The overall profile and the magnitude of the velocity are similar to those obtained in the Lagrangian averaging (Figure 7b).

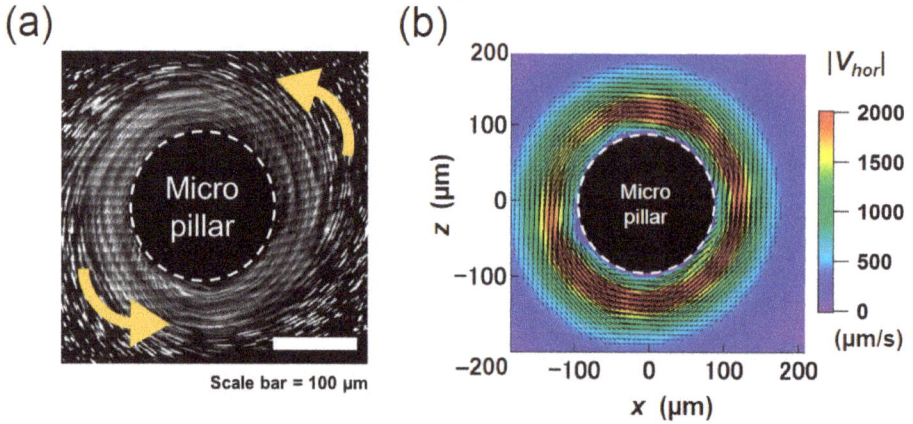

Figure 8. (a) Raw image of fluorescent tracer beads overlaid for sixty cycles. (b) 2D vector plot of the averaged velocity flow field obtained from PIV measurement.

5.2. Comparison of Radial Velocity Profile

To compare the results quantitatively, we plotted the radial profiles of $|V_{hor}|$ in Figure 9. The profiles were averaged in the azimuthal direction of the pillar. The magnitude of the peak velocity obtained by Eulerian method is twice as large as the other, and the peak position of the velocity is closer to the pillar, as qualitatively seen in Figure 7. On the other hand, the velocity distributions obtained by the Lagrangian method and PIV measurement showed a similar trend in terms of the peak position (r = 130 μm; 30 μm away from the pillar wall) and the decay of the profile. There is a difference between the two profiles close to the pillar; this difference could be partly caused by the difficulty in resolving the high-shear region in both simulation and PIV. The size of the single grid in the simulation corresponds to be 8.3 μm, and the size of the window for image correlation in PIV corresponds to be 6.4 μm. Particle images tended to blur in the region close to the wall due to the 3D flow described later. Nonetheless, a good agreement in the decaying profile after the peak supports the validity of the present simulation.

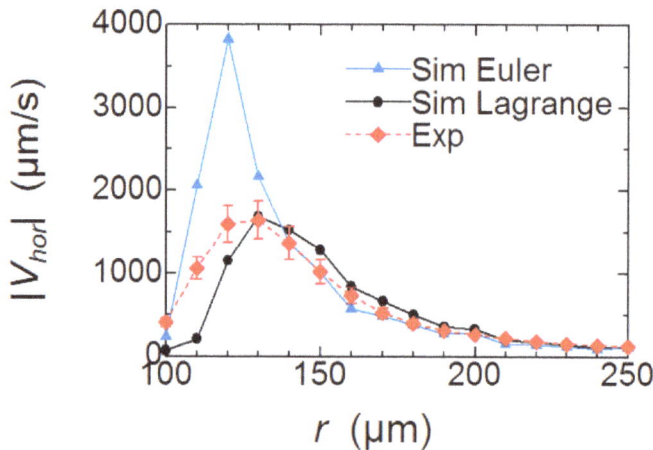

Figure 9. Distribution of the mean horizontal velocity magnitude of particles in radial direction. (▲) Eulerian averaging and (●) Lanrangian averaging of the simulation result. (◆) PIV measurement. PIV was repeated three times with different setup, and the average and standard deviation are shown.

Overall, the significant difference between two averaging methods is reasonable considering that the flow is unsteady. The virtual fluid particles in the flow do not stay at the constant radial position r with respect to the center of the pillar. Instead, their radial position changes as they rotated around the reference position (center of the rotational path) at each phase. Thus, the velocity at the actual particle position is different from that at the reference position. As this slight difference accumulates, the time averaged velocity per one cycle becomes different from the average velocity obtained at the stationary coordinates. The present result confirms that it is necessary to use the Lagrangian tracking method to reproduce the unsteady flow around the vibrating pillar.

5.3. Comparison of Normal Velocity Profile

Next, we examined the velocity distribution in the vertical plane including the Stokes layer. To do this we conducted the micro PIV measurement from $y = 0$ to 100 µm at 10 µm interval. The plane at $y > 100$ µm was not obtained experimentally due to the limitation in the working distance of the objective. The color maps of $|V_{hor}|$ in r-y plane obtained in the numerical simulation (Lagrangian averaging) and the PIV measurement are shown in Figure 10a,b, respectively. They showed a similar trend in terms of the peak position of $r = 130$ µm at $40 < y < 80$ µm, which slightly moves closer to the pillar near the top ($y \sim 100$ µm). This is attributed to the strong three-dimensionality of the flow around the tip of the pillar, as will be shown later. The magnitude of the velocity decays rapidly as the position gets closer to the bottom wall.

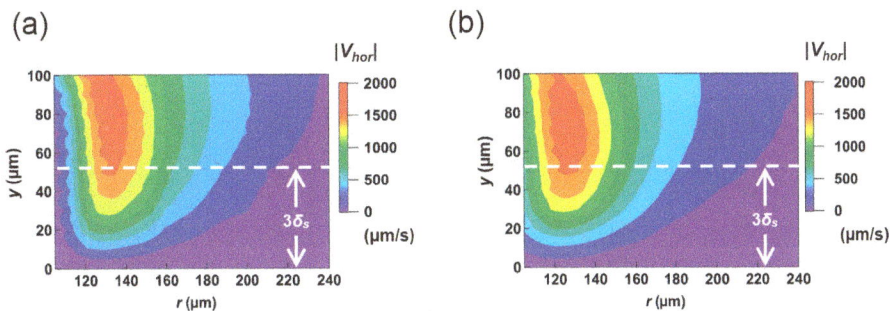

Figure 10. Contour plots of mean horizontal velocity magnitude $|V_{hor}|$ in r-y plane of (**a**) numerical result (Lagrangian averaging) and (**b**) PIV measurement. The dashed line indicates the Stokes layer thickness calculated based on the present experimental condition.

The horizontal velocity distributions in the vertical (y-) direction at $r = 130$ µm are plotted shown in Figure 11. Both profiles agreed well, with velocity values close to 0 µm/s on the substrate surface ($y = 0$ µm), which increased to ~1.8 mm/s towards the tip of the pillar. The peak value is obtained in the vicinity of $y = 70$ to 80 µm.

In the Stokes second problem of a unidirectional oscillation of the flat solid wall in contact with the fluid, the fluid in the vicinity of the wall almost follows the wall movement, but this influence exponentially decays as it moves away from the wall with a characteristic length $\delta_s = (2\nu/\omega)^{1/2}$ [13]. From its analytical solution, the effect of the wall motion decreases down to ~5% ($1/e^3$) at $y = 3\delta_s$. In the present condition, $3\delta_s$ corresponds to 53.4 µm by substituting $\nu = 1.0 \times 10^{-6}$ m^2/s (25 °C) and $\omega = 2\pi f = 6.28 \times 10^3$ rad/s. The height is indicated by dotted lines in Figures 10 and 11. Since the net flow discussed in the present system is induced by the relative velocity between the fluid and the substrate, the maximum average flow field occurs outside the Stokes layer. In contrast, the fluid within the Stokes layer thickness $3\delta_s$ is dragged by the motion of the wall due to the viscous effect so that the relative velocity diminishes on the bottom wall.

The present results indicate that the viscous effect adjacent to the solid wall has a strong impact on the net travel distance of a passive tracer, and the thickness of this fluid layer is determined by the

Stokes layer thickness. It should be also noted that, for each height y, the peak of the averaged velocity occurs about 30 μm away from the side wall of the pillar as shown in Figure 10. This horizontal gap is also similar to the Stokes layer thickness. When SS is applied to micro devices, the Stokes layer thickness is not negligibly small in general, and this causes the strong dependency of the induced averaged flow on the vertical location, i.e., the distance from the top and bottom substrate. Therefore, in order to predict and design the entire velocity field inside a micro device, 3D analysis is requisite.

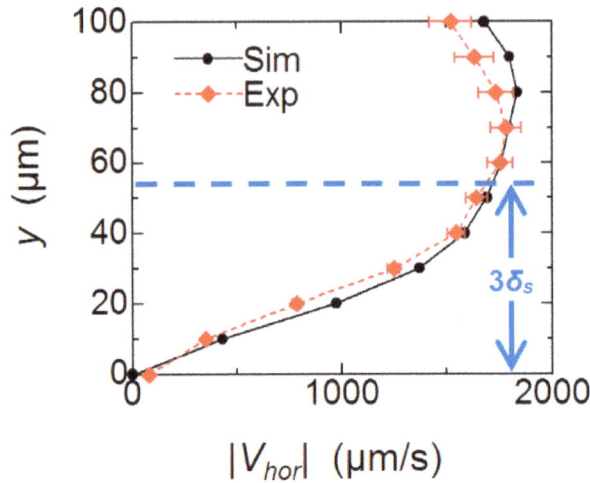

Figure 11. Comparison of vertical distributions of mean horizontal velocity magnitude $|V_{hor}|$.

5.4. Three-Dimensionality of the Flow

Lastly, we investigated the three-dimensionality of the averaged velocity field around the pillar. Figure 12 shows a vector and a contour plot of the averaged velocity in the vertical plane ($|V_{ver}| = \sqrt{u_1^2 + u_2^2}$) obtained from the numerical simulation. In the vicinity of the upper and lower wall surfaces, it is clear that there is almost no upward/downward motion, so that the flow is practically two-dimensional. However, at the intermediate region between the top and bottom plates, the velocity toward the pillar with the magnitude as large as ~200 μm/s appears around the corner of the pillar ($r = 130$ to 150 μm and $y = 50$ to 100 μm). This flow is diverted upward at the pillar edge. As a result, a vortical motion is generated close to the corner of the pillar. Although the magnitude is about 10 times smaller than $|V_{hor}|$, the presence of the vertical velocity component has a strong impact on the averaged velocity field as discussed in Figure 10.

To confirm the three-dimensionality of the averaged velocity flow field in the experimental flow field, the motion of the particle was observed from the side of the pillar using an objective lens with a long working distance placed horizontally. Because the confocal illumination is not possible away from the wall, we introduced 10 μm polystyrene beads as tracers into the fluid at a low concentration so that individual beads were visible from a long distance. Two representative paths visualized by superimposing ~50 successive frames (1.7 s) are shown in the left column in Figure 13a,b (see Movie S2 and S3 for corresponding movies). In both cases, tracer particles exhibited three-dimensional motions including upward and downward movements, instead of a simple orbital movement in the same horizontal plane. In Figure 13a, a bead initially circulating around the middle of the pillar was suddenly raised at the region close the apex, and then hovered above the pillar. The similar path could be reproduced by tracking an ideal tracer in the averaged velocity field obtained by the Lagrangian averaging (Figures 7b and 12) in the simulation (right figures). This sudden ascension should be caused by the upward flow near the apex discussed in Figure 12. Another bead shown in Figure 13b, which was circulating a bit far outside of the pillar slightly, shifted downward as it got closer to the pillar, and then raised. This path was also reproduced in the numerical simulation (right figures), which was convected by the downward net velocity at $r = 120 \sim 170$ μm and $y = 100 \sim 130$ μm, and raised by the upward net velocity at $r = 110 \sim 150$ μm and $y = 40 \sim 100$ μm shown in Figure 12. These trajectories show the validity of numerical prediction for three-dimensionality of flow induced around the pillar at the present vibration condition.

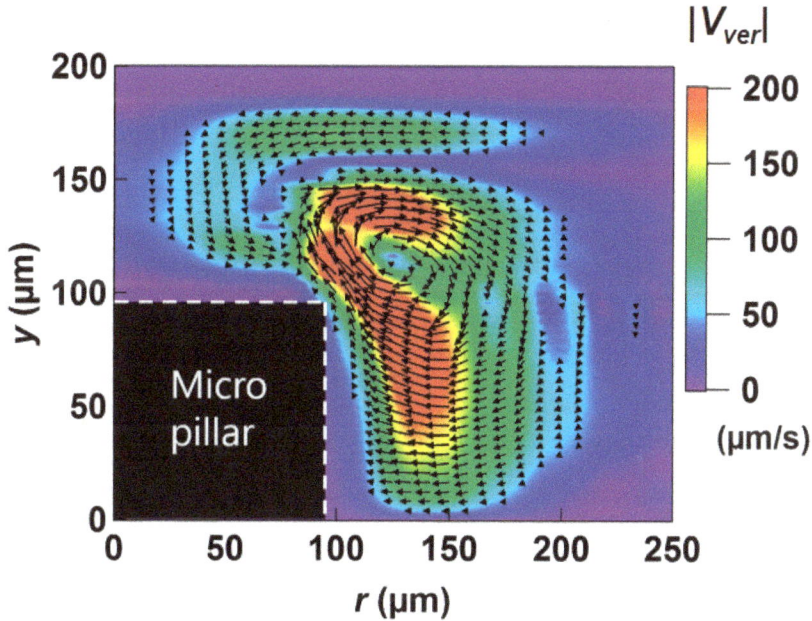

Figure 12. Vector plot of the averaged velocity flow (Lagrangian method) in the vertical plane.

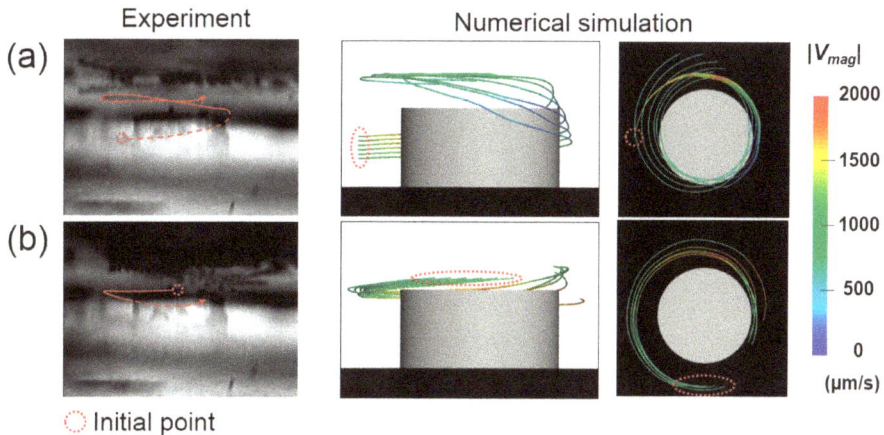

Figure 13. Comparison of particle trajectories obtained in the experiment and the numerical simulation. (**a**) Ascending motion near the apex of the pillar. (**b**) Descending and ascending motion.

6. Conclusions

In this work, we developed a simulation tool to predict the SS flow induced by vibration without assuming 2D flow and small vibration amplitude. The developed numerical code was based on the volume penalization method in the Cartesian coordinate system, so that arbitrary three-dimensional structures can be readily embedded without requiring grid generation for each geometry. As pointed out in the past SS studies, we confirmed that the average velocity at fixed points (Eulerian mean) and that of fluid particles advected by the local flow velocity (Lagrangian mean) are essentially different. In practical microdevice applications, the Lagrangian-averaged flow governs mass transport and mixing. In this work, we compared the Lagrangian mean field obtained by numerical calculation with micro PIV experimental results. The quantitative agreement between them validates the present simulation method.

Through these numerical and experimental analyses, we observed two 3D characteristics of the SS flow even at the flow Reynolds number of 10. First, inside the Stokes layer, which develops from the solid surface toward the fluid region, the averaged velocity due to the Stokes drift diminishes because the fluid is dragged by the solid motion due to the viscous effect, and thereby the relative velocity between the fluid and the solid reduces. Since the thickness of the Stokes layer cannot be neglected with respect to the length scale of microdevices in general, it is necessary to accurately predict the influence of the Stokes layer in order to understand the flow inside the device. Secondly, a large strong vortex motion accompanying upward and downward motions was confirmed in the vicinity of the apex of the micro-pillar. Such vertical fluid motions could have significant influences on mass transport, and may result in a complicated flow field combined with the three-dimensionality caused by the Stokes layer mentioned above. Since the present numerical results reproduce the 3D nature of the flow fields observed in experiment, our numerical code could be a powerful tool to optimize the structure of a micro-pillar and the vibration mode in order to develop innovative microfluidic devices in future work.

Supplementary Materials: The following are available online at http://www.mdpi.com/2072-666X/9/12/668/s1, Supplementary Text: Governing equations of fluid in a moving frame with rotational vibration. Figure S1: Comparison with the radial profile of $|V_{hor}|$ calculated with different grid resolutions in the numerical simulation. The distribution calculated with a low-resolution grid ($48 \times 25 \times 48$) differ significantly, but those calculated with medium ($96 \times 49 \times 96$; used in the main results) and high-resolutions ($144 \times 73 \times 144$) were similar., Figure S2: Comparison of the radial profile of $|V_{hor}|$ calculated with different window size in PIV analysis. One pixel in images corresponds to 0.8 μm. Supporting Movie 1: Animation of the velocity field (u_1, u_3) at $y = 50$ μm obtained from the numerical simulation., Supporting Movie 2: Motion of a tracer particle depicted in Figure 13a in the main text., Supporting Movie 3: Motion of a tracer particle depicted in Figure 13b in the main text.

Author Contributions: H.S. and Y.H. conceived the project. K.K., Y.K., Y.H., and H.S. designed and conducted the numerical simulation. K.K. and T.O. designed and conducted the PIV measurement. K.K. and T.H. performed the horizontal visualization. K.K., Y.H., and H.S. wrote the paper and all authors revised and approved the manuscript.

Funding: This research was supported by the Chuo University Grant for Special Research and Tokyo Ohka Foundation for the promotion of science and technology. Y.H. gratefully acknowledges the support by the Ministry of Education, Culture, Sports, Science and Technology of Japan (MEXT) through the Grant-in-Aid for Scientific Research (B) (No. 17H03170).

Acknowledgments: We thank Akira Fukawa for initial exploration of the project. We also thank Toshiyuki Matsui and Yuta Kobayashi for helping the horizontal visualization experiment.

Conflicts of Interest: The authors declare no conflict of interest.

References

1. Wiklund, M.; Green, R.; Ohlin, M. Acoustofluidics 14: Applications of acoustic streaming in microfluidic devices. *Lab Chip* **2012**, *12*, 2438–2451. [CrossRef] [PubMed]
2. Karimi, A.; Yazdi, S.; Ardekani, A.M. Hydrodynamic mechanisms of cell and particle trapping in microfluidics. *Biomicrofluidics* **2013**, *7*, 021501. [CrossRef] [PubMed]
3. Hagiwara, M.; Kawahara, T.; Arai, F. Local streamline generation by mechanical oscillation in a microfluidic chip for noncontact cell manipulations. *Appl. Phys. Lett.* **2012**, *101*, 074102. [CrossRef]

4. Huang, P.H.; Xie, Y.L.; Ahmed, D.; Rufo, J.; Nama, N.; Chen, Y.C.; Chan, C.Y.; Huang, T.J. An acoustofluidic micromixer based on oscillating sidewall sharp-edges. *Lab Chip* **2013**, *13*, 3847–3852. [CrossRef] [PubMed]

5. Ohlin, M.; Christakou, A.E.; Frisk, T.; Onfelt, B.; Wiklund, M. Influence of acoustic streaming on ultrasonic particle manipulation in a 100-well ring-transducer microplate. *J. Micromech. Microeng.* **2013**, *23*, 035008. [CrossRef]

6. Huang, P.H.; Nama, N.; Mao, Z.M.; Li, P.; Rufo, J.; Chen, Y.C.; Xie, Y.L.; Wei, C.H.; Wang, L.; Huang, T.J. A reliable and programmable acoustofluidic pump powered by oscillating sharp-edge structures. *Lab Chip* **2014**, *14*, 4319–4323. [CrossRef] [PubMed]

7. Hayakawa, T.; Akita, Y.; Arai, F. Parallel trapping of single motile cells based on vibration-induced flow. *Microfluid. Nanofluid.* **2018**, *22*, 42. [CrossRef]

8. Hayakawa, T.; Sakuma, S.; Arai, F. On-chip 3D rotation of oocyte based on a vibration-induced local whirling flow. *Microsyst. Nanoeng.* **2015**, *1*, 15001. [CrossRef]

9. Hayakawa, T.; Sakuma, S.; Fukuhara, T.; Yokoyama, Y.; Arai, F. A Single Cell Extraction Chip Using Vibration-Induced Whirling Flow and a Thermo-Responsive Gel Pattern. *Micromachines* **2014**, *5*, 681–696. [CrossRef]

10. Ahmed, D.; Ozcelik, A.; Bojanala, N.; Nama, N.; Upadhyay, A.; Chen, Y.; Hanna-Rose, W.; Huang, T.J. Rotational manipulation of single cells and organisms using acoustic waves. *Nat. Commun.* **2016**, *7*, 11085. [CrossRef]

11. Ozcelik, A.; Nama, N.; Huang, P.H.; Kaynak, M.; McReynolds, M.R.; Hanna-Rose, W.; Huang, T.J. Acoustofluidic Rotational Manipulation of Cells and Organisms Using Oscillating Solid Structures. *Small* **2016**, *12*, 5120–5125. [CrossRef] [PubMed]

12. Kim, E.; Kojima, M.; Xiaoming, L.; Hattori, T.; Kamiyama, K.; Mae, Y.; Arai, T. Analysis of rotational flow generated by circular motion of an end effector for 3D micromanipulation. *ROBOMECH J.* **2017**, *4*, 5. [CrossRef]

13. Schlichting, H.; Gersten, K. *Boundary-Layer Theory*, 8th ed.; Springer: New York, NY, USA, 2000.

14. Amin, N.; Riley, N. Streaming from a Sphere Due to a Pulsating Source. *J. Fluid Mech.* **1990**, *210*, 459–473. [CrossRef]

15. Riley, N. Steady streaming. *Annu. Rev. Fluid Mech.* **2001**, *33*, 43–65. [CrossRef]

16. Lieu, V.H.; House, T.A.; Schwartz, D.T. Hydrodynamic Tweezers: Impact of Design Geometry on Flow and Microparticle Trapping. *Anal. Chem.* **2012**, *84*, 1963–1968. [CrossRef] [PubMed]

17. House, T.A.; Lieu, V.H.; Schwartz, D.T. A model for inertial particle trapping locations in hydrodynamic tweezers arrays. *J. Micromech. Microeng.* **2014**, *24*, 045019. [CrossRef]

18. Nama, N.; Huang, P.H.; Huang, T.J.; Costanzo, F. Investigation of acoustic streaming patterns around oscillating sharp edges. *Lab Chip* **2014**, *14*, 2824–2836. [CrossRef] [PubMed]

19. Nama, N.; Huang, P.H.; Huang, T.J.; Costanzo, F. Investigation of micromixing by acoustically oscillated sharp-edges. *Biomicrofluidics* **2016**, *10*, 024124. [CrossRef]

20. Nama, N.; Huang, T.J.; Costanzo, F. Acoustic streaming: an arbitrary Lagrangian-Eulerian perspective. *J. Fluid Mech.* **2017**, *825*, 600–630. [CrossRef]

21. Lutz, B.R.; Chen, J.; Schwartz, D.T. Microscopic steady streaming eddies created around short cylinders in a channel: Flow visualization and Stokes layer scaling. *Phys. Fluids* **2005**, *17*, 023601. [CrossRef]

22. Amit, R.; Abadi, A.; Kosa, G. Characterization of steady streaming for a particle manipulation system. *Biomed. Microdevices* **2016**, *18*. [CrossRef] [PubMed]

23. Rallabandi, B.; Marin, A.; Rossi, M.; Kahler, C.J.; Hilgenfeldt, S. Three-dimensional streaming flow in confined geometries. *J. Fluid Mech.* **2015**, *777*, 408–429. [CrossRef]

24. Hasegawa, Y.; Kasagi, N. Dissimilar control of momentum and heat transfer in a fully developed turbulent channel flow. *J. Fluid Mech.* **2011**, *683*, 57–93. [CrossRef]

25. Osher, S.; Sethian, J.A. Fronts Propagating with Curvature-Dependent Speed—Algorithms Based on Hamilton-Jacobi Formulations. *J. Comput. Phys.* **1988**, *79*, 12–49. [CrossRef]

26. Kolomenskiy, D.; Schneider, K. A Fourier spectral method for the Navier-Stokes equations with volume penalization for moving solid obstacles. *J. Comput. Phys.* **2009**, *228*, 5687–5709. [CrossRef]

27. Suzuki, T.; Hasegawa, Y. Estimation of turbulent channel flow at Re_τ =100 based on the wall measurement using a simple sequential approach. *J. Fluid Mech.* **2017**, *830*, 760–796. [CrossRef]

28. Suzuki, H.; Ho, C.M.; Kasagi, N. A chaotic mixer for magnetic bead-based micro cell sorter. *J. Microelectromech. Syst.* **2004**, *13*, 779–790. [CrossRef]

micromachines

MDPI

Article

Evaluation of Lipid Accumulation Using Electrical Impedance Measurement under Three-Dimensional Culture Condition

Daiki Zemmyo [1] and Shogo Miyata [2],*

[1]   Graduate School of Science and Technology, Keio University, 3-14-1 Hiyoshi, Yokohama 223-8522, Japan
[2]   Department of Mechanical Engineering, Faculty of Science and Technology, Keio University, 3-14-1 Hiyoshi, Yokohama 223-8522, Japan
*   Correspondence: miyata@mech.keio.ac.jp; Tel.: +81-45-566-1827

Received: 17 May 2019; Accepted: 2 July 2019; Published: 6 July 2019

Abstract: The degeneration of adipocyte has been reported to cause obesity, metabolic syndrome, and other diseases. To treat these diseases, an effective *in vitro* evaluation and drug-screening system for adipocyte culture is required. The objective of this study is to establish an *in vitro* three-dimensional cell culture system to enable the monitoring of lipid accumulation by measuring electrical impedance, and to determine the relationship between the impedance and lipid accumulation of adipocytes cultured three dimensionally. Consequently, pre-adipocytes, 3T3-L1 cells, were cultured and differentiated to the adipocytes in our culture system, and the electrical impedance of the three-dimensional adipocyte culture at a high frequency was related to the lipid accumulation of the adipocytes. In conclusion, the lipid accumulation of adipocytes could be evaluated in real time by monitoring the electrical impedance during *in vitro* culture.

Keywords: electrical impedance measurement; three-dimensional cell culture; adipocyte; lipid droplet; 3T3-L1

---

1. Introduction

Lipids are a critical factor for maintaining cellular energy homeostasis. Energy is stored as tryglycerides in lipid droplets when additional energy is ingested and hydrolyzed into fatty acids during energy shortage [1]. Although lipids contribute to the survival of living organisms, they cause diseases such as obesity and metabolic syndrome, both of which are related to lipid accumulation. Obesity, which is defined as an increase in adipose tissue, is a major problem worldwide, and considered to be caused by the intake of high calorie foods. Lysosomal diseases (LDs) are a type of genetic disease that cause metabolic disorders and are characterized by the accumulation of byproducts in the lysosome, owing to a defective catabolism. LDs are designated as an orphan disease [2] and occur in 1 out of 8000 live births [3]; therefore, a therapeutic approach is required to treat LDs. To establish medical treatment for various diseases related to metabolism effectively, it is essential to establish *in vitro* evaluation systems for lipid accumulation.

In adipocyte activity research, many types of assays have been used to evaluate the processes of adipogenesis and lipid accumulation. For example, Oil Red O staining is a primary evaluation method for the accumulation of lipid droplets in adipocytes [4,5]. Furthermore, glycerol-3-phosphate dehydrogenase activity [6], triglyceride content, and marker of genes such as PPARγ have been measured to evaluate adipocyte metabolism. However, these assays are only end-point assays and do not enable the real-time monitoring of adipocyte activities. A histological assessment is required to obtain photomicrographs such that they can be analyzed to evaluate the degree of stained area.

Therefore, it is essential to develop a new technique to evaluate the degree of lipid accumulation quantitatively and conveniently in real time.

Electric cell-substrate impedance sensing was established by Giaever and Keese in 1984 [7]. It is a technique for measuring cellular properties, such as cell density [8], cell aging [9], cell adhesion [10], and cytotoxicity [11]. Furthermore, this sensing technology has been applied in monitoring cell differentiation, for example, stem cell differentiation [12] and neural differentiation [13]. It can also be applied in cell monitoring without cellular damage and hence enable a real-time evaluation. Because adipose tissue exhibits higher electrical impedance than other tissues, it is considered that the electrical characteristics of adipocytes would change during lipid accumulation. Several reports have been published regarding the monitoring of electrical characteristics during lipid accumulation in adipocytes [14,15]. However, few studies have evaluated the relationship between the electrical characteristics and the amount of lipid droplets in adipocytes. Furthermore, the evaluation was not performed under physiological conditions, because the measurement was only performed in the two-dimensional culture of adipocytes. Therefore, a platform for a screening system or a fundamental study to evaluate lipid accumulation under a three-dimensional culture is required.

In this study, we developed a cell culture device that can simultaneously measure electrical characteristics under a three-dimensional cell culture condition. Moreover, the relationship between the electrical characteristics and the lipid accumulation of adipocytes cultured three-dimensionally was determined using our novel cell culture device.

2. Materials and Methods

2.1. Impedance Measurement of Three-Dimensional Cell Culture

An impedance measurement device was developed to evaluate the electrical characteristics of adipocytes cultured under three-dimensional conditions (Figure 1). The device was composed of two platinum wires bridged between two polycarbonate fixtures in each well of a six-well cell culture plate. A cell-seeded collagen gel disk could be cultured in each well. The two platinum wires were embedded in the gel and held by two polycarbonate fixtures to be positioned in parallel with each other (10 mm apart) and 1 mm height from the bottom surface of the six-well plate. The length of each platinum wire in the cell-seeded gel was set as 20 mm, and the diameter was 0.2 mm. Each platinum wire passed through the inside of the polycarbonate fixture and was connected to the electrical impedance meter (Chemical Impedance Meter, Hioki, Japan). The complex impedance between two platinum wires was measured to evaluate the electrical characteristics of the three-dimensionally cultured adipocytes. Furthermore, this device could be set in a CO_2 gas incubator to enable real-time impedance monitoring during cell culture.

To evaluate how to pass electric currents between the two platinum wires, numerical analysis was performed to evaluate the electric field in the cultured area of the device, using finite element analysis software (COMSOL Multiphysics Version 5.2a, COMSOL Inc., Stockholm, Sweden). In the analysis, the conductivity and relative permittivity were set as follows: 1.38 S/m [16–18] and 80 [16] in culture medium; 0.30 S/m [19] and 2.28 [20] in collagen gel. The result of this numerical analysis indicated that electric currents passed through both areas of culture medium and collagen gel (Figure 2). The proportion of electric current that flowed in the collagen gel was approximately 20% of the total electric current between the two wires and was considered as sufficient to reflect the electrical characteristics of the cell-seeded collagen gel.

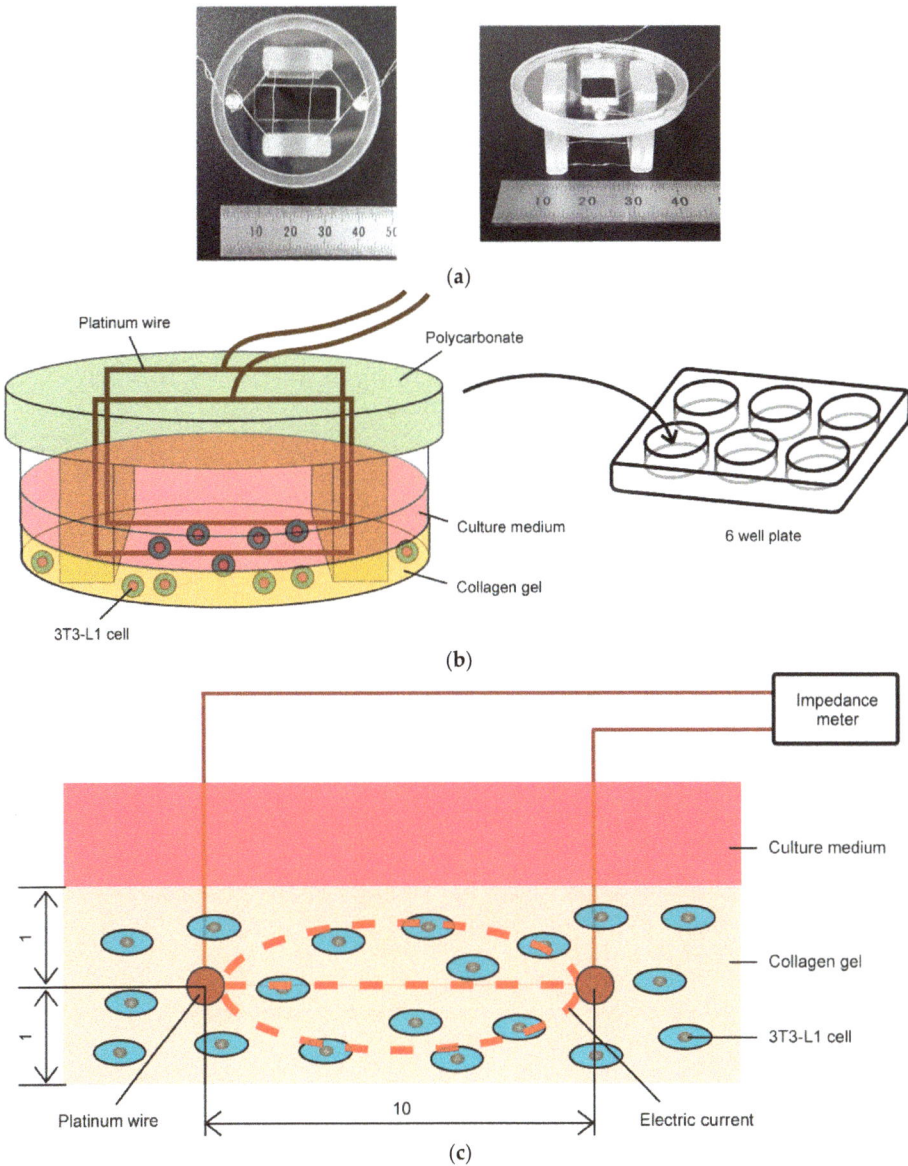

Figure 1. (**a**) Cell culture device for electrical impedance measurement. (**b**) Two platinum wires were fixed with a polycarbonate fixture and placed inside the 3T3-L1-cell-embedded collagen gel. It was placed within a 6-well cell-culture plate. (**c**) Cross-sectional view of the culture region of impedance measurement device. Voltage was applied between the two platinum wires.

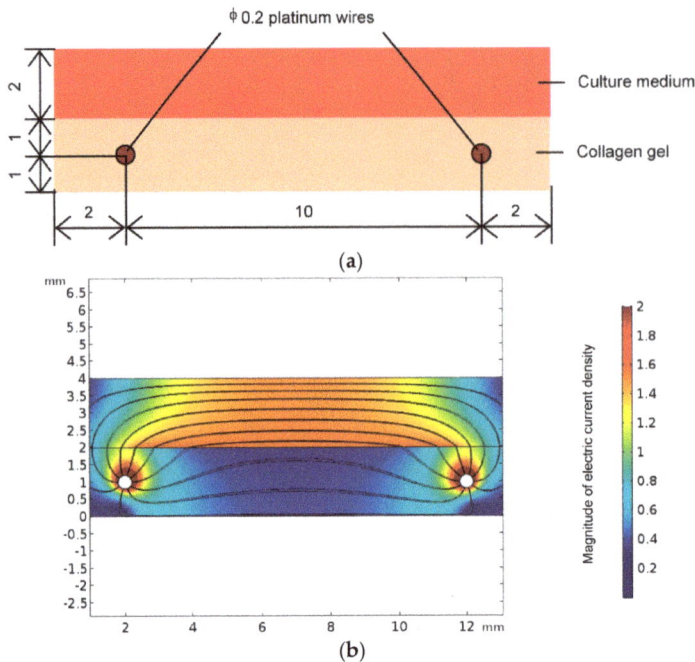

Figure 2. (**a**) Numerical analysis model of electrical field. (**b**) Pathway of electric currents between two platinum wires in the cell culture region. Streamlines show the pathway of the electric current, and the contour plot shows the magnitude of the electric current density.

2.2. Evaluation of Characteristics of Living Cell

Intracellular and extracellular properties can be evaluated separately by measuring the electrical characteristics of cells in multiple frequency bands. Ions in a cytoplasm are gathered around the surface of a cell membrane to increase the apparent dielectric constant of cytoplasm if the cells are subjected to a low-frequency alternating-current (AC) electric field (Figure 2). The electric currents can pass through the surface of the cell membrane but not through the cytoplasm because of the increased dielectric constant of the cytoplasm. By contrast, if cells are subjected to a high-frequency AC electric field, the ions inside the cells cannot follow the change in electric field to decrease the dielectric constant of the cytoplasm. Furthermore, the apparent electrical impedance decreases in relation to the decrease in dielectric properties of the cytoplasm to enable the penetration of electric currents into the cell membrane. Therefore, if a high-frequency AC electric field is applied to living cells, the measured data would reflect the electrical characteristics of intracellular constituents.

As for the two-dimensional culture of pre-adipocytes (3T3-L1 cells) and adipocytes, it was reported that the effect of lipid accumulation on the change in impedance at lower frequencies (10–65 kHz) was negligible [15]. Meanwhile, the impedance at higher frequencies (1–15 MHz) changed according to the lipid accumulation [21]. These results indicate that lipid accumulation could not affect the impedance at lower frequencies (approximately 10 kHz) but could affect the impedance at higher frequencies (1 MHz). The impedance at lower frequencies could be related partially to the electrical characteristic of cell membranes and primarily to the experimental setup of the measurement device. Therefore, we propose an impedance at a higher frequency corrected by a lower frequency, i.e., Z(1 MHz)' = Z(1 MHz) − Z(10 kHz) as an evaluation parameter for the change in intracellular constituents, especially for the lipid accumulation.

2.3. Cell Culture with Impedance Measurement

Murine pre-adipocyte 3T3-L1 cells were maintained in 75 cm^2 flasks in Dulbecco's modified eagle medium (DMEM, high glucose, Gibco), supplemented with 10% Equa-FETAL (Equa FETAL, EF-0500-A, Atlas Biologicals, Fort Collins, CO, USA) and 2% antibiotic–antimycotic (Antibiotic-Antimycotic Mixed Stock Solution, Nacalai Tesque, Japan) in a humidified CO$_2$ incubator (5% CO$_2$ at 37 °C). From a cryopreserved stock, 3T3-L1 cells were passaged twice before being embedded in collagen gel. In this study, two types of basal mediums were used for 3T3-L1 cells to evaluate the effect of glucose concentration on lipid accumulation. The lipid metabolism is closely related to the glucose metabolism in adipocyte. DMEM with high glucose (4.5 g/L) was used for the high-glucose condition (high-G group), and DMEM with low glucose (1.0 g/L) was used for the low-glucose condition (low-G group). The basal medium supplemented with 1.0 μM dexamethasone, 0.50 mM isobutyl-methylxanthine, and 10 μg/mL insulin was prepared as an adipocyte differentiation medium. A basal medium with 10 μg/mL insulin was prepared for the maintenance culture after adipocyte differentiation. A collagen gel without 3T3-L1 cells was prepared as a no-cell group to confirm the effect of 3T3-L1 cells on the electrical impedance.

In the high-G and low-G groups, the 3T3-L1 cells were suspended with 2.4 mg/mL neutralized type-I collagen solution at a concentration of 1.0×10^6 cells/mL. A 2.0 mL of cell-suspended collagen solution was poured into the well of the impedance measurement device and gelled for 30 min at 37 °C in the incubator. In the no-cell group, the neutralized type-I collagen solution without cells was poured into the well of the device, following gelation in the incubator. An amount of 2.0 mL of basal medium was poured into the well after the gelation, and the cell-seeded collagen gel was cultured for 48 h. Subsequently, the culture medium was changed to a differentiation medium to induce the adipogenesis of 3T3-L1 cells. Following adipogenesis for 48 h, the culture medium was changed to a maintenance medium. In this experiment, at the start of the maintenance culture, the time was set as t = 0 h (Figure 3). The 3T3-L1 cells in the collagen gel were cultured until t = 288 h. The culture medium was changed every 48 h during the maintenance culture.

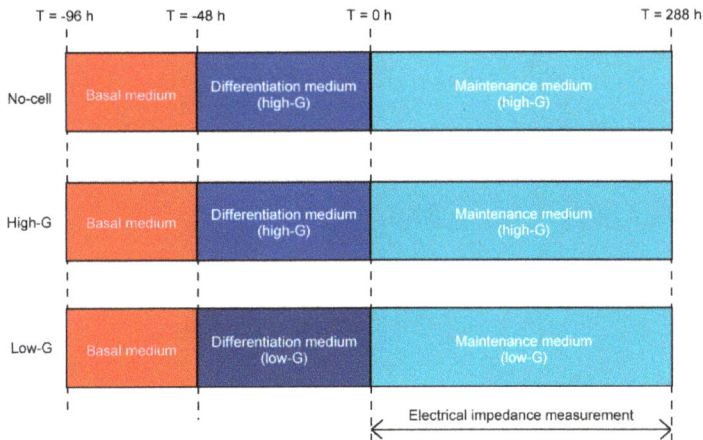

Figure 3. Time-course of cell culture experiments. The 3T3-L1 cells were cultured three-dimensionally in basal medium for 48 h following adipogenesis in differentiation medium for 48 h. After adipogenesis (t = 0 h), the differentiated 3T3-L1 cells were cultured in maintenance medium and the electrical impedance of the cells was measured until t = 288 h.

2.4. Biochemical Characterization

In this study, to focus on real-time and in situ monitoring of lipid accumulation, the evaluation of lipid accumulation was performed without fluorescent staining of cells. The lipid accumulation was evaluated from the phase-contrast images of the cells. The images were acquired at the center plane between the top and bottom surface of cell-seeded gel. In our preliminary studies, the lipids in 3T3-L1 cells were stained by Oil red O and demonstrated sphere-like shapes (Figure S1) [22]. Based on these results, the circle-like constituents from 1 μm to 10 μm in the 3T3-L1 cells were defined as lipids, and the total area of these constituents was measured for the evaluation of lipid accumulation.

To examine the effect of cell mass on electrical characteristics in collagen gel, the cell number was evaluated by the quantification of total DNA amount in collagen gel. It was reported that the total DNA was related to the cell number in hydrogels or living tissues [23,24]. After the cell culture experiment, the samples were lyophilized overnight and treated with 125 μg/mL papain solution at 60 °C for 6 h to solubilize the collagen gel. The total DNA amount in the digested specimen was determined using a fluorescence spectrophotometer (Qubit 2.0 Fluorometer, Life Technologies, Carlsbad, CA, USA).

2.5. Statistical Analysis

Most of the data are representative of three individual experiments with similar results. For each group, 4 samples (n = 4) were analyzed per time point, and each data point represents the mean and standard deviation. Data from each experimental group were examined for significant differences using t-tests. Pearson's correlation analyses were also carried out on pooled data sets of high-G and low-G groups to determine the relationship between the electrical impedance and lipid accumulation.

3. Results and Discussion

3.1. Effect of Glucose Concentration on Proliferation and Lipid Accumulation of 3T3-L1 Cells

In this study, to evaluate the number of living cells in collagen gel at the end of culture time, the amount of total DNA in the cultured specimen was quantified using a fluorometric assay. The quantification of total DNA was used for the evaluation of cell number in living tissues or three-dimensional cultures [25]. No significant differences were found between the DNA amount of high-G and low-G groups (Figure 4). Therefore, it was assumed that the change in electrical impedance was primarily dependent on the lipid accumulation and adipogenesis of 3T3-L1 cells.

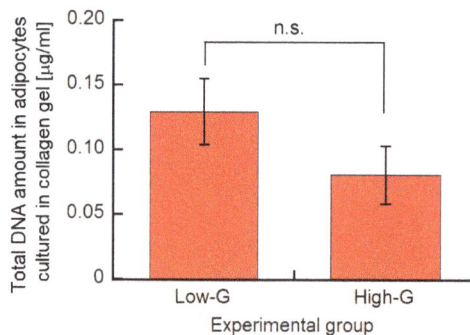

Figure 4. Total DNA amount in adipocytes cultured in collagen gel. Data are presented as mean ± S.D., n = 4.

For the evaluation of lipid accumulation in 3T3-L1 cells, the phase-contrast images of the cells in the collagen gel were acquired every 48 h. Before adipocyte differentiation, at t = −48 h, the cells demonstrated similar morphology and the cell number was similar in both the high-G and low-G groups (Figure 5a). At t = 96 h, small lipid droplets were observed in the cells in both groups (Figure 5b); therefore, it was revealed that the 3T3-L1 cells in collagen gel differentiated into adipocytes and accumulated lipid droplets. With increasing culture time, the number of lipid droplets increased in both groups and the size of each droplet became larger (Figure 5c,d). Therefore, it was suggested that the 3T3-L1 pre-adipocytes could be cultured three-dimensionally and the amount of lipid droplets could increase with culture time.

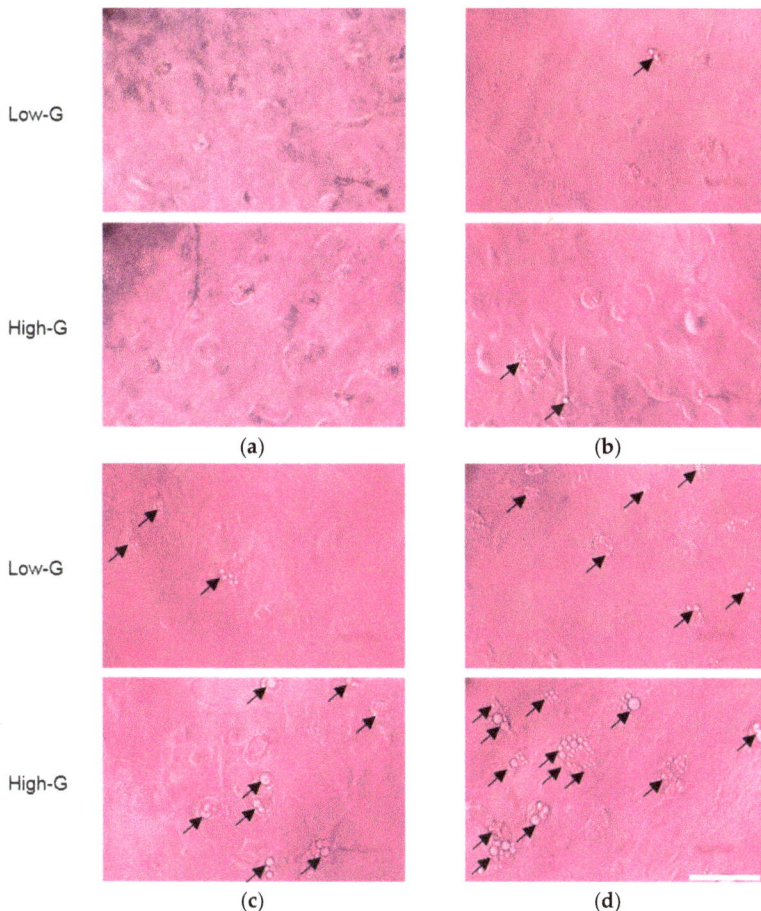

Figure 5. Phase-contrast images of 3T3-L1 cells cultured in collagen gel at (**a**) −48 h, (**b**) 96 h, (**c**) 192 h, and (**d**) 288 h post adipogenesis. Lipid droplets were accumulated after t = 96 h (black arrows indicate lipid droplets). Scale bar: 50 μm.

The area ratio of lipid droplets to the entire imaged area at every 48 h is shown in Figure 6. During the adipocyte differentiation period (from t = −48 h to 0 h), no lipid droplets were detected in both the high-G and low-G groups, whereas a small number of lipids was observed at t = 48 h. In the low-G group, the area ratio of the lipid droplet increased at a rate of approximately 0.5% per 48 h until t = 192 h and did not change after t = 192 h. Meanwhile, in the high-G group, the area ratio increased at a rate of approximately 0.5% per 48 h until t = 192 h and at a rate of approximately 1% per 48 h after t = 192 h. At each time point, the proportion of lipid droplets in the high-G group was larger than that in the low-G group and a significant difference was observed after t = 96 h. Therefore, it was considered that the concentration of glucose in the culture medium would affect the accumulation of lipid droplets in adipocytes.

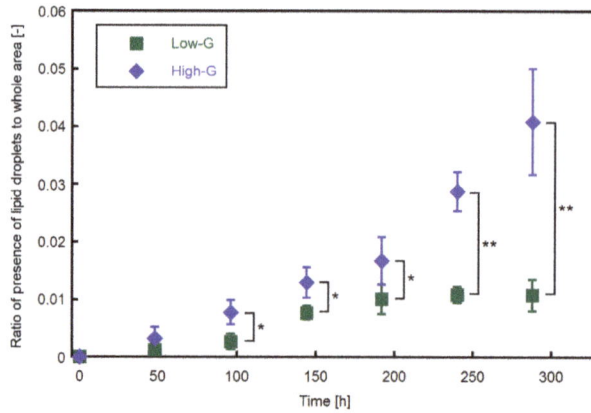

Figure 6. Area proportion of lipid droplets in phase-contrast microscopic images. The amount of lipid droplets in the high-G group was significantly larger than that in the low-G group after t = 96 h. Data are presented as mean ± S.D., n = 4. * and ** indicate significant differences between low-G and high-G groups (*: $p < 0.05$, **: $p < 0.01$).

3.2. Electrical Impedance of 3T3-L1 Cells Cultured Three-Dimensionally

As described in the "Materials and Methods" section, the corrected impedance at higher frequencies, Z(1 MHz)' = Z(1 MHz) − Z(10 kHz), was defined as the evaluation of the electrical impedance. The change in Z(1 MHz)' from t = 0 h, Z(1 MHz)$'_t$ − Z(1 MHz)$'_0$, is shown in Figure 7. In the no-cell group, the Z(1 MHz)' did not change during the culture time. Meanwhile, the Z_{1MHz}' of the low-G group decreased monotonically and that of the high-G group decreased until t = 144 h and saturated from t = 144 h to 288 h.

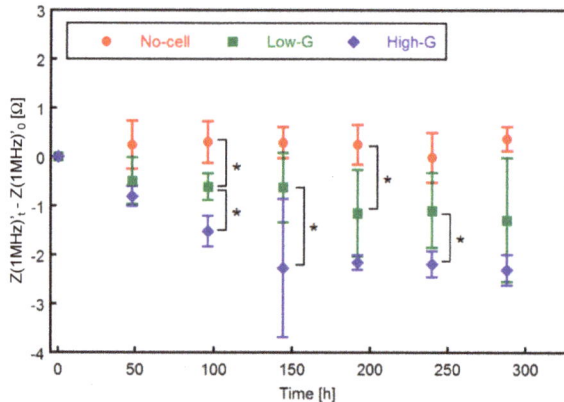

Figure 7. Change in Z(1 MHz)$'_t$ − Z(1 MHz)$'_0$ in three experimental groups. Both the values of low and high-G groups decreased with increase in culture time. Data are presented as mean ± S.D., n = 4. * indicates a significant difference between low-G and high-G groups, $p < 0.05$.

Significant differences between the Z(1 MHz)$'_t$ − Z(1 MHz)$'_0$ values of the no-cell and cell-seeded groups (high-G and low-G groups) were detected at t = 96 h and 192 h. Moreover, a significant difference between the values of low-G and high-G groups was also observed from 96 h to 144 h and 240 h. In previous studies, the electric property of a living cell was modeled as an equivalent circuit consisting of resistances and capacitance (Figure 8) [25]. At higher frequencies, the dielectric constant of the inner cell constituents would decrease with lipid accumulation, thus resulting in the decrease in conductance. Therefore, the impedance at higher frequencies was considered to decrease with lipid accumulation. Our experimental results were consistent with the analysis based on the equivalent circuit model of a cell. As mentioned in Section 3.1, no significant difference was found between the cell numbers of the high-G and low-G groups. The difference in the values of Z(1 MHz)$'_t$ − Z(1 MHz)$'_0$ could be related to lipid accumulation in the cell.

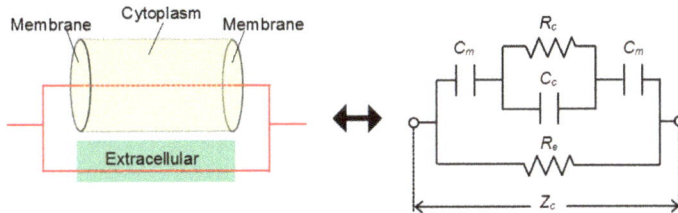

Figure 8. Equivalent circuit of a living cell. Cytoplasm was modeled as a parallel circuit of resistance R_c and capacitance C_c, cell membrane as capacitance C_m, and extracellular medium as resistance R_e.

3.3. Relationship between Lipid Accumulation and Electrical Impedance

The relationship between the time change in electrical impedance at high frequencies, Z(1 MHz)$'_t$ − Z(1 MHz)$'_0$ and the area ratio of lipid droplets to the total area of phase-contrast images is shown in Figure 9. Significant negative correlations were found between the change in Z(1 MHz)$'_t$ − Z(1 MHz)$'_0$ and the area proportion of the lipid droplets ($R^2 = 0.55$, $p < 0.05$ for low-G group; $R^2 = 0.65$, $p < 0.05$ for high-G group). These experimental results were qualitatively consistent with the equivalent circuit model of a cell (Figure 8) [25]. It was suggested that the amount of lipid droplets in a three-dimensional adipocyte culture could be evaluated by monitoring the change in impedance at high frequencies, Z(1 MHz)$'$ during an *in vitro* culture.

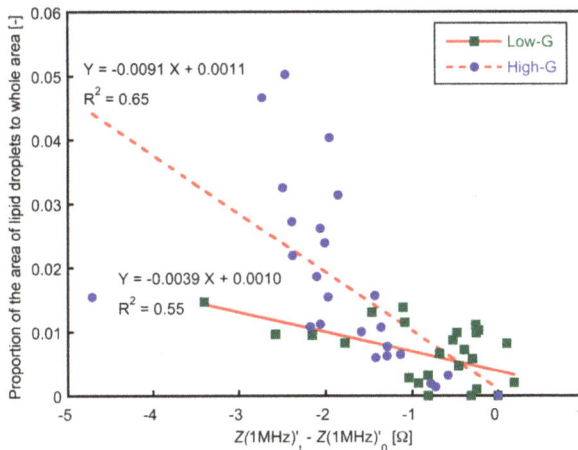

Figure 9. The relationship between Z(1 MHz)$'_t$ − Z(1 MHz)$'_0$ and the proportion of area of lipid droplets to entire area. The entire data in low-G and high-G groups are plotted. A significant negative correlation was found ($R^2 = 0.49$, $p < 0.05$).

4. Conclusions

In this study, a cell culture device that could simultaneously measure electrical characteristics under a three-dimensional culture was developed. The pre-adipocytes, i.e., 3T3-L1 cells, were cultured and differentiated in collagen gel and their electric property was evaluated during the culture. The relationship between the amount of lipid droplets and the electrical impedance of three-dimensionally cultured adipocytes was evaluated. Results indicated that the lipid accumulation of adipocytes could be evaluated in real time by measuring the impedance at higher frequencies.

Supplementary Materials: The following are available online at http://www.mdpi.com/2072-666X/10/7/455/s1, Figure S1: (**a**) Phase-contrast and (**b**) Oil red O stained images of differentiated 3T3-L1 cells. Scale bar: 100 μm.

Author Contributions: Conceptualization, D.Z. and S.M.; methodology, D.Z.; validation, D.Z. and S.M.; investigation, D.Z.; resources, S.M.; data curation, D.Z.; writing—original draft preparation, D.Z.; writing—review and editing, S.M.; visualization, D.Z.; supervision, S.M.; project administration, S.M.; funding acquisition, S.M.

Funding: This research was partially supported by of JSPS KAKENHI (Grant numbers: 17K01369 and 26560222) and the Translational Research Network Program from Japan Agency for Medical Research and Development (AMED).

Conflicts of Interest: The authors declare no conflict of interest.

References

1. Singh, R.; Kaushik, S.; Wang, Y.; Xiang, Y.; Novak, I.; Komatsu, M.; Tanaka, K.; Cuervo, A.M.; Czaja, M.J. Autophagy regulates lipid metabolism. *Nature* **2009**, *458*, 1131–1135. [CrossRef] [PubMed]
2. Platt, F.M. Emptying the stores: lysosomal diseases and therapeutic strategies. *Nat. Rev. Drug Discov.* **2017**, *17*, 133–150. [CrossRef] [PubMed]
3. Smith, J.; Harm, T.; Levine, D.; Christopher, S.; Hostetter, S.; Snella, E.M.; Johnson, G.; Ellinwood, N.M. Novel lysosomal disease in a juvenile golden retriever. *Mol. Genet. Metab.* **2018**, *123*, S133. [CrossRef]
4. Escorcia, W.; Ruter, D.L.; Nhan, J.; Curran, S.P. Quantification of Lipid Abundance and Evaluation of Lipid Distribution in Caenorhabditis elegans by Nile Red and Oil Red O Staining. *J. Vis. Exp.* **2018**, *133*, e57352. [CrossRef] [PubMed]
5. Koopman, R.; Schaart, G.; Hesselink, M.K. Optimisation of oil red O staining permits combination with immunofluorescence and automated quantification of lipids. *Histochem. Cell Biol.* **2001**, *116*, 63–68. [PubMed]
6. Chyau, C.-C.; Chu, C.-C.; Chen, S.-Y.; Duh, P.-D. The inhibitory effects of djulis (chenopodium formosanum) and its bioactive compounds on adipogenesis in 3T3-L1 adipocytes. *Molecules* **2018**, *23*, 1780. [CrossRef] [PubMed]
7. Giaever, I.; Keese, C.R. Monitoring fibroblast behavior in tissue culture with an applied electric field. *Proc. Natl. Acad. Sci. USA* **1984**, *81*, 3761–3764. [CrossRef] [PubMed]
8. Angstmann, M.; Brinkmann, I.; Bieback, K.; Breitkreutz, D.; Maercker, C. Monitoring human mesenchymal stromal cell differentiation by electrochemical impedance sensing. *Cytotherapy* **2011**, *13*, 1074–1089. [CrossRef]
9. Jun, H.-S.; Dao, L.T.M.; Pyun, J.-C.; Cho, S. Effect of cell senescence on the impedance measurement of adipose tissue-derived stem cells. *Enzym. Microb. Technol.* **2013**, *53*, 302–306. [CrossRef]
10. Xiao, C.; Lachance, B.; Sunahara, G.; Luong, J.H.T. An In-Depth Analysis of Electric Cell−Substrate Impedance Sensing to Study the Attachment and Spreading of Mammalian Cells. *Anal. Chem.* **2002**, *74*, 1333–1339. [CrossRef]
11. Asphahani, F.; Zhang, M. Cellular Impedance Biosensors for Drug Screening and Toxin Detection. *Analyst* **2007**, *132*, 835–841. [CrossRef] [PubMed]
12. Zhou, Y.; Basu, S.; Laue, E.; Seshia, A.A. Single cell studies of mouse embryonic stem cell (mESC) differentiation by electrical impedance measurements in a microfluidic device. *Biosens. Bioelectron.* **2016**, *81*, 249–258. [CrossRef] [PubMed]
13. Park, H.E.; Kim, D.; Koh, H.S.; Cho, S.; Sung, J.-S.; Kim, J.Y. Real-Time Monitoring of Neural Differentiation of Human Mesenchymal Stem Cells by Electric Cell-Substrate Impedance Sensing. *J. Biomed. Biotechnol.* **2011**, *2011*, 1–8. [CrossRef] [PubMed]
14. Lee, R.; Jung, I.; Park, M.; Ha, H.; Yoo, K.H. Real-time monitoring of adipocyte differentiation using a capacitance sensor array. *Lab Chip* **2013**, *13*, 3410–3416. [CrossRef] [PubMed]
15. Bagnaninchi, P.O.; Drummond, N. Real-time label-free monitoring of adipose-derived stem cell differentiation with electric cell-substrate impedance sensing. *Proc. Natl. Acad. Sci. USA* **2011**, *108*, 6462–6467. [CrossRef] [PubMed]

16. Fuhr, G.; Müller, T.; Schnelle, T.; Hagedorn, R.; Voigt, A.; Fiedler, S.; Arnold, W.M.; Zimmermann, U.; Wagner, B.; Heuberger, A. Radio-frequency microtools for particle and liver cell manipulation. *Naturwissenschaften* **1994**, *81*, 528–535. [CrossRef] [PubMed]

17. Fuhr, G.; Shirley, S.G. Cell handling and characterization using micron and submicron electrode arrays: state of the art and perspectives of semiconductor microtools. *J. Micromech. Microeng.* **1995**, *5*, 77–85. [CrossRef]

18. Puttaswamy, S.V.; Sivashankar, S.; Chen, R.-J.; Chin, C.-K.; Chang, H.-Y.; Liu, C.H.; Chen, R.; Chin, C.; Chang, H. Enhanced cell viability and cell adhesion using low conductivity medium for negative dielectrophoretic cell patterning. *Biotechnol. J.* **2010**, *5*, 1005–1015. [CrossRef]

19. Macdonald, R.A.; Voge, C.M.; Kariolis, M.; Stegemann, J.P. Carbon nanotubes increase the electrical conductivity of fibroblast-seeded collagen hydrogels. *Acta Biomater.* **2008**, *4*, 1583–1592. [CrossRef]

20. Da Cruz, A.G.; Góes, J.; Figueiró, S.; Feitosa, J.P.; Ricardo, N.M.P.; Sombra, A.S. On the piezoelectricity of collagen/natural rubber blend films. *Eur. Polym. J.* **2003**, *39*, 1267–1272. [CrossRef]

21. Schade-Kampmann, G.; Huwiler, A.; Hebeisen, M.; Hessler, T.; Di Berardino, M. On-chip non-invasive and label-free cell discrimination by impedance spectroscopy. *Cell Prolif.* **2008**, *41*, 830–840. [CrossRef] [PubMed]

22. Aulthouse, A.L.; Freeh, E.; Newstead, S.; Stockert, A.L. Part 1: A Novel Model for Three-Dimensional Culture of 3T3-L1 Preadipocytes Stimulates Spontaneous Cell Differentiation Independent of Chemical Induction Typically Required in Monolayer. *Nutr. Metab. Insights* **2019**, *12*, 117863881984139. [CrossRef] [PubMed]

23. Kim, Y.-J.; Sah, R.L.; Doong, J.-Y.H.; Grodzinsky, A.J. Fluorometric assay of DNA in cartilage explants using Hoechst. *Anal. Biochem.* **1988**, *174*, 168–176. [CrossRef]

24. Forsey, R.W.; Chaudhuri, J.B. Validity of DNA analysis to determine cell numbers in tissue engineering scaffolds. *Biotechnol. Lett.* **2009**, *31*, 819–823. [CrossRef] [PubMed]

25. Hernández-Balaguera, E.; López-Dolado, E.; Polo, J.L. Obtaining electrical equivalent circuits of biological tissues using the current interruption method, circuit theory and fractional calculus. *RSC Adv.* **2016**, *6*, 22312–22319. [CrossRef]

micromachines

MDPI

Article

Effect of Cyclic Stretch on Tissue Maturation in Myoblast-Laden Hydrogel Fibers

Shinako Bansai [1,†], Takashi Morikura [1,†], Hiroaki Onoe [2] and Shogo Miyata [2,*]

[1] Graduate School of Science and Technology, Keio University, 3-14-1 Hiyoshi, Yokohama 223-8522, Japan; shinako0220@keio.jp (S.B.); dnngu-1elife@keio.jp (T.M.)
[2] Department of Mechanical Engineering, Faculty of Science and Technology, Keio University, 3-14-1 Hiyoshi, Yokohama 223-8522, Japan; onoe@mech.keio.ac.jp
* Correspondence: miyata@mech.keio.ac.jp; Tel.: +81-45-566-1827
† These authors have equally contributed to this work.

Received: 17 May 2019; Accepted: 13 June 2019; Published: 15 June 2019

Abstract: Engineering of the skeletal muscles has attracted attention for the restoration of damaged muscles from myopathy, injury, and extraction of malignant tumors. Reconstructing a three-dimensional muscle using living cells could be a promising approach. However, the regenerated tissue exhibits a weak construction force due to the insufficient tissue maturation. The purpose of this study is to establish the reconstruction system for the skeletal muscle. We used a cell-laden core-shell hydrogel microfiber as a three-dimensional culture to control the cellular orientation. Moreover, to mature the muscle tissue in the microfiber, we also developed a custom-made culture device for imposing cyclic stretch stimulation using a motorized stage and the fiber-grab system. As a result, the directions of the myotubes were oriented and the mature myotubes could be formed by cyclic stretch stimulation.

Keywords: myoblast; skeletal muscle; core-shell hydrogel fiber; cyclic stretch; engineered muscle

1. Introduction

Muscle tissue consists of an ordered muscle fiber array, which is tightly bundled, long, and cylindrical multinucleated myotube cells. Muscles play an important role in daily human activities including metabolic regulation of internal organs. Myopathy, injury, and extraction of malignant tumors are some of the common issues to restore muscle tissue. Therefore, the tissue engineering approach for muscle regeneration is beneficial. There are several approaches to reconstruct the three-dimensional muscle tissue: Cell-seeded collagen gel [1], cell-based sheets [2], cell aggregates [3], etc. In addition, aligned electrospun nanofibers are also used for scaffold materials, to promote cellular alignment [4,5]. These approaches also have the potential for in vitro drug screening and disease modeling [6–9]. Muscle tissue can be regenerated by these approaches. However, the maturation of the engineered tissue is still far from the "native" muscle [10].

In this study, we control the orientation of myoblast-like cells and reconstruct the maturated muscle tissues by mechanical stimuli. The effect of mechanical stimulation on cell homeostasis and development, which are critical factors in tissue maintenance, repair, and regeneration, has drawn a lot of attention. As mechanical stimuli to promote tissue regeneration, the stimuli mimicking in vivo physiological condition was imposed on living cells: Shear stress or stretch for blood vessel remodeling [11,12], stretch for the bone [13] and ligament [14] remodeling, and stretch or electric stimuli [15,16] for muscle remodeling. For myogenesis, the mechanical and chemical stimulations promote the myogenesis of myoblasts or myoblast-like cells to become multinucleated myotube cells [17]. It has been reported that mechanical stretch can affect the remodeling of the cytoskeleton in myocytes [18,19]. Considering the results of previous studies, mechanical stretch was used as the stimuli in this study. Nguyen et al. already developed the cell culture device to impose cyclic stimuli on a cell-seeded sheet-shaped scaffold and reported the effect of mechanical stimuli on fibrous tissue reconstruction [20]. However, their research was performed regarding the fibroblast culture on the sheet-shaped scaffold, which was not suitable to simulate muscle tissue. Skeletal muscle has the possibility to restore itself after minor injury. However, promotion of myogenesis by mechanical stimuli also benefits cardiac muscle tissue engineering and has drawn a lot of attention in severe cardiac disease cases.

To mimic the "native" muscle structure, we focused on cell fiber technology [21]. This technology encapsulates living cells into the core region of a hydrogel core-shell microfiber, allowing the cells to grow, migrate, promote cell-cell interaction, and form a fiber-shaped tissue called "cell fiber". Using this cell fiber technology based on the hydrogel tube structure, gases (O_2 and CO_2) and nutrients are allowed to penetrate into the core region containing cells [22], leading to an efficient cell expansion with high viability.

Here, we develop a custom-made culture device for "cell fiber" to impose mechanical stretch cyclically on the cell fiber using a motorized stage and the fiber-grab system. In addition, we also evaluate the effect of the cyclic stretch on in vitro skeletal muscle regeneration.

2. Materials and Methods

2.1. Cells

Mature murine myogenic cell line C2C12 cells were purchased from Riken Cell Bank (Tsukuba, Japan). The culture medium was Dulbecco's modified essential medium (DMEM, Sigma, St. Louis, MO, USA) containing 10% fetal bovine serum (FBS), and 1% antibiotic/antimycotic solution (A/A, Thermo Fisher Scientific, Waltham, MA, USA). The cells were maintained in a 5% CO_2 atmosphere at 37 °C in a CO_2 incubator and used for experiments before they reached 5 passages.

2.2. Formation of Core-Shell Hydrogel Microfibers

According to previous studies, C2C12 cells were cultured in collagen gel [23,24]. To encapsulate C2C12 cells suspended in the collagen gel in the core region of alginate fibers, the double-coaxial laminar-flow microfluidic device was fabricated by assembling pulled glass capillary tubes, rectangular glass tubes, and custom-made three-way connectors, as previously described (Figure 1) [16]. Three solutions were required for core-shell hydrogel microfiber formation: (1) core stream: A solution of C2C12 cells suspended in 4.0 mg/mL neutralized type I collagen (AteloCell®, IC-50, KOKEN, Tokyo, Japan) at 1.8×10^8 cells/mL, (2) shell stream: A solution of 1.5 wt % sodium alginate (80–120 cP, Wako Pure Chemical Industries, Osaka, Japan), and (3) sheath stream: A solution of 100 mM calcium chloride ($CaCl_2$, Kanto Chemicals, Tokyo, Japan) with 3% w/w sucrose (Nacalai Tesque, Kyoto, Japan). The flow rates of the core, shell, and sheath streams were 25 µL/min, 120 µL/min, and 3.6 mL/min, respectively. The fabricated fibers were finally cultured in the culture medium for 24 h to induce the collagen gelation and for cell adhesion.

The differentiation protocol for C2C12 cells were already standardized. However, that for the three-dimensional culture of C2C12 cells were not fully established. To validate the culture medium for three-dimensional culture of C2C12 cells, three types of medium, DMEM with 2% horse serum (HS) [25], 10% HS, and 10% FBS, were tested for preliminary study. As a result, the cells in fibers cultured in DMEM with 2% and 10% HS tended to decrease whereas the cells cultured in DMEM with 10% FBS tended to increase in a 6-day culture (Figure 2). Based on this data, DMEM with 10% FBS was determined as the culture medium for C2C12-cell fibers.

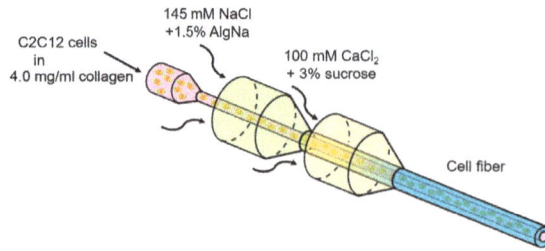

Figure 1. Schematic for fabrication of core-shell hydrogel microfibers. The C2C12 cell-laden core-shell hydrogel microfiber was formed by the double co-axial laminar flow.

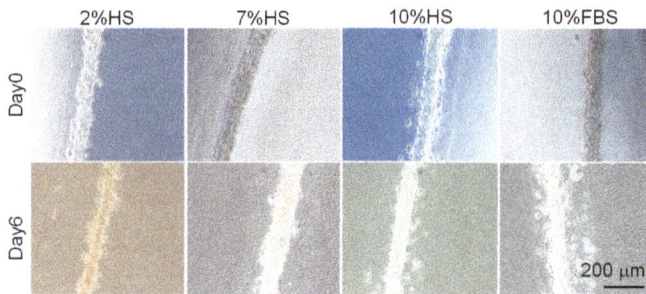

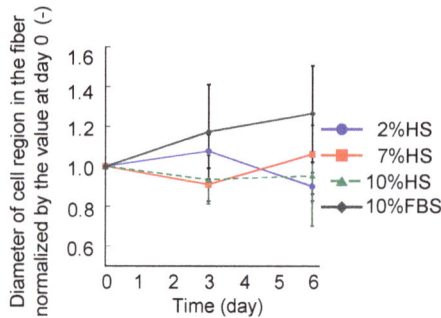

Figure 2. Cell-laden core-shell hydrogel microfiber culture under 2%, 7%, and 10% horse serum (HS) and 10% fetal bovine serum (FBS) conditions. (**a**) Phase-contrast images and (**b**) the change in the diameter of cell-laden core along with culture time. Scale bar: 200 μm.

2.3. Three-Dimensional Cell Culture with Cyclic Stretch

To impose cyclic stretch on the cell fibers, a custom-made stretching device was developed. The device was composed of a motorized stage and a culture chamber containing two guide rods to hold the cell fiber (Figure 3a). The cell fibers were wrapped around two parallel rods to stretch the cell fibers and the distance of the rods was changed cyclically using a computer controlled motorized stage (Figure 3b). Briefly, the guide rods were set to be parallel (10 mm apart) using a supporting block and the fibers were wrapped around the rods. After the fiber wrapping, 4.0 mg/mL type I collagen solution was dropped on the connecting part of the fibers and the guide rods to ensure the adhesion. Following collagen gelation, the rods were removed and connected to the custom-made stretching device (Figure 3c). The chamber was then filled with 5 mL culture medium to immerse the fibers in the medium. The stretching device was set in a CO_2 incubator to culture the fibers in a 5% CO_2 atmosphere at 37 °C. After 2-day static culture, the cell fibers were subjected to 3% tensile strain at 1 Hz for 4 h/day for 2 days. The tensile strain and frequency were decided according to previous studies to avoid the destruction of hydrogel fibers [26]. For control specimens, the cell fibers were cultured under same condition except for the cyclic stretch.

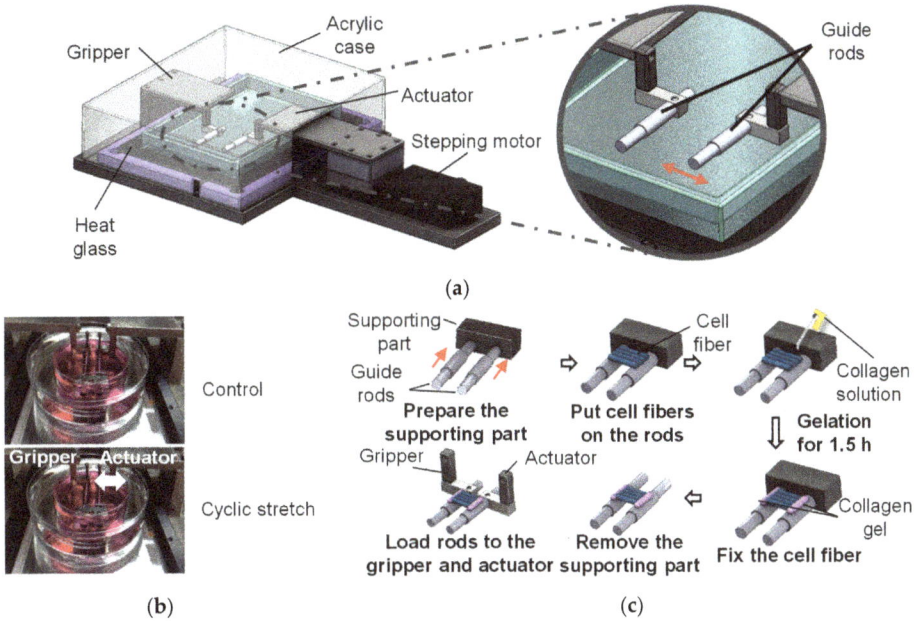

Figure 3. Custom-made cell culture device for "cell fiber" to impose the cyclic stretch. (**a**) Schematic of the culture device, (**b**) photograph of the gripper for hydrogel fibers, and (**c**) procedure to grab the fibers using the collagen gel and two stainless rods.

2.4. Microscopy and Image-Based Analysis

To evaluate the myogenesis of C2C12 cells, phase-contrast images were acquired after the 4-day culture (2-day static culture following a 2-day cyclically stretch stimulation). The cells in hydrogel fibers were also stained with calcein-AM to evaluate the morphology of live cells, with rhodamine-phalloidin to evaluate the cytoskeleton. The calcein-AM stains cytoplasm of live cells and the rhodamine phalloidin stains actin filaments. For the calcein-AM staining, the fibers were firstly washed with a serum-free medium two times and incubated with 0.1 mg/mL calcein-AM in DMEM for 30 min. For the rhodamine-phalloidin staining, the cells in the fiber were fixed with 4% paraformaldehyde for 10 min following permeabilization with 0.1% Triton X-100 in phosphate buffered saline (PBS) for 5 min at room temperature. After cell fixation, the cell fibers were incubated with 0.7% rhodamine-phalloidin (PHDR1, Cytoskeleton) for 30 min at 37 °C. After fluorescent staining, the cells were observed by a fluorescent microscope (CKX41, Olympus, Tokyo, Japan) equipped with a CCD camera (DP73, Olympus) and a confocal scanning microscope (FV10i-DOC, Olympus).

To evaluate the myogenesis of C2C12 cells from fluorescent images, the image-based analysis was performed using Image J software (NIH). The fluorescent images were preprocessed using a smooth filter and a sharpen filter with 3×3 neighborhood. After the preprocessing, the images were converted to 8-bit grayscale images and binarized using Otsu's method. Finally, the cell regions in the binary images were fitted to ellipses and the aspect ratio of each ellipse was measured. In this study, cultured

C2C12 cells were divided into three groups: (1) undifferentiated cells (aspect ratio < 2.0), (2) immature myotube-like cells (2.0 ≤ aspect ratio < 3.0), and (3) mature myotube-like cells (aspect ratio ≥ 3.0).

3. Results and Discussion

3.1. Difference in Tissue Remodeling in the Cell Fibers and in the Two-Dimensional Culture

To evaluate the effect of the three-dimensional culture condition on myogenesis of C2C12 cells, the cytoskeletons of both the monolayer culture and the cell fiber were evaluated (Figure 4). The direction of the cytoskeleton in the monolayer culture was random, whereas the cytoskeleton in the cell fibers aligned to the cylindrical axis of the fiber. It was suggested that the C2C12 cells reorganized their structure of cytoskeleton to align the wall of the gel fiber. Many studies reported that the direction of the cells aligned to the groove of the culture substrate to reorganize the cytoskeleton of the cell [27–30]. The result of our study was consistent with these studies.

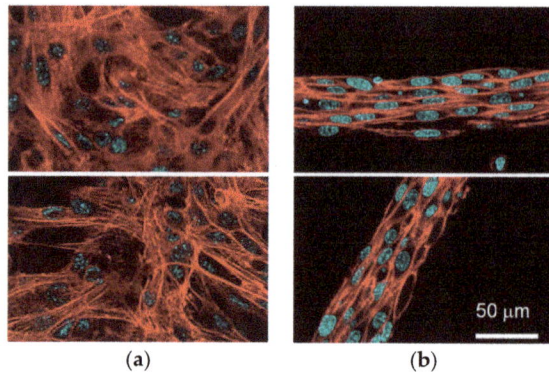

(a) **(b)**

Figure 4. Fluorescent images of rhodamine-phalloidine/DAPI counterstaining to visualize the actin cytoskeleton of the C2C12 cells. (**a**) Monolayer culture and (**b**) three-dimensional culture using a hydrogel microfiber culture of the C2C12 cells. Scale bar: 50 μm.

3.2. Effect of Cyclic Stretch on Tissue-Reconstruction in the Cell Fibers

The diameter of the C2C12-cell region in the cell fibers subjected to the cyclic stretch decreased as compared to that in the control group (Figure 5a). Using calcein-AM staining, almost all the cells were positively stained in both the cyclic-stretch and the control group (Figure 5b). This result indicates that the cell viability was maintained in our custom-made cell culture device. The cells in the cyclic-stretch group elongated themselves and aligned to the axis of the fiber. The cells in the control group, on the other hand, were uniformly distributed and did not elongate themselves. As shown in Figure 6, the actin cytoskeleton of the cells in the cyclic-stretch group was concentrated and also aligned to the axis of the cell fiber as compared to the ones in the control group. In this study, we assessed myogenesis of the C2C12 cells based on the aspect ratio of each cell (Figure 7). Cyclic stretch promoted the myogenesis of the C2C12 cells and increased the ratio of the mature myotube-like cells as compared to the ones in the control group. Approximately 70% of the cells were differentiated in the cyclic-stretch group whereas approximately 50% of the cells were differentiated in the control group. Moreover, the ratio of the mature myotube-like cells in the cyclic stretch group was over two times larger than that of the cells in the control group.

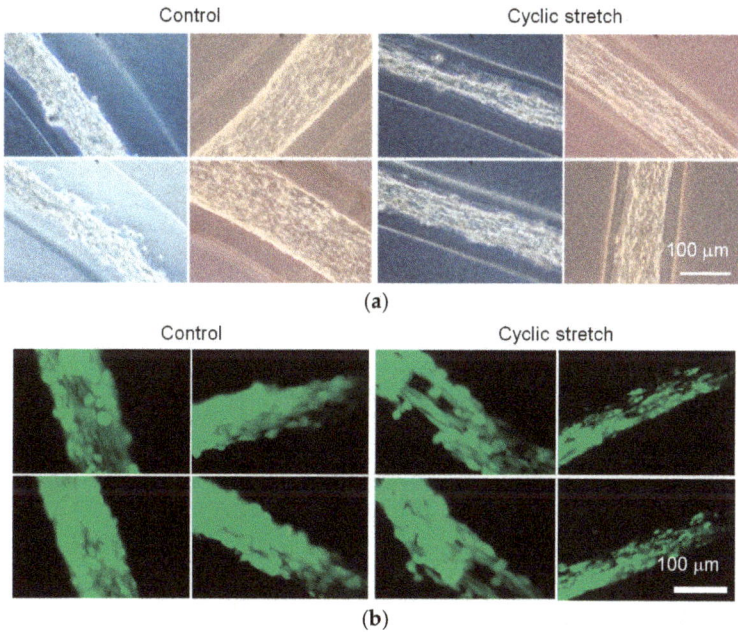

Figure 5. (**a**) Phase-contrast and (**b**) fluorescent images (calcein-AM staining) of C2C12-cell laden hydrogel fibers subjected to cyclic stretch. Scale bar: 100 μm.

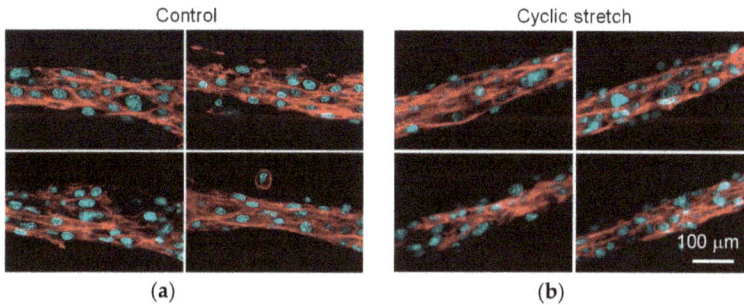

Figure 6. Fluorescent images of rhodamine-phalloidine/DAPI counterstaining to visualize the actin cytoskeleton of C2C12 cells in (**a**) control and (**b**) cyclic stretch groups. Scale bar: 100 μm.

In order to reconstruct the skeletal muscle tissue for the tissue engineering therapy, it is important to culture the cells three-dimensionally with physical stimuli. For in vitro skeletal muscle regeneration, various physical factors are reported to align the cells and progress tissue maturation [15,16,27–31]. Among them, mechanical stress-like tension or electrical stimuli have been reported to affect the cell alignment and maturation in vitro [15,16]. For myoblasts or myoblast-like cells, the stretch could enhance the myosin expression to promote myogenesis [26]. Consistent with these studies, it was considered that the cyclic stretch promoted the myogenesis of the C2C12 cells and the maturation of the muscle fibers in cell-laden hydrogel fibers. Especially, cells in the skeletal muscles, also in addition to the cardiac muscles, are constantly subjected to cyclic mechanical stretch to generate highly differentiated and maturated muscle fibers. Therefore, mechanical stimuli could be an important factor for tissue regeneration of the skeletal and the cardiac muscles.

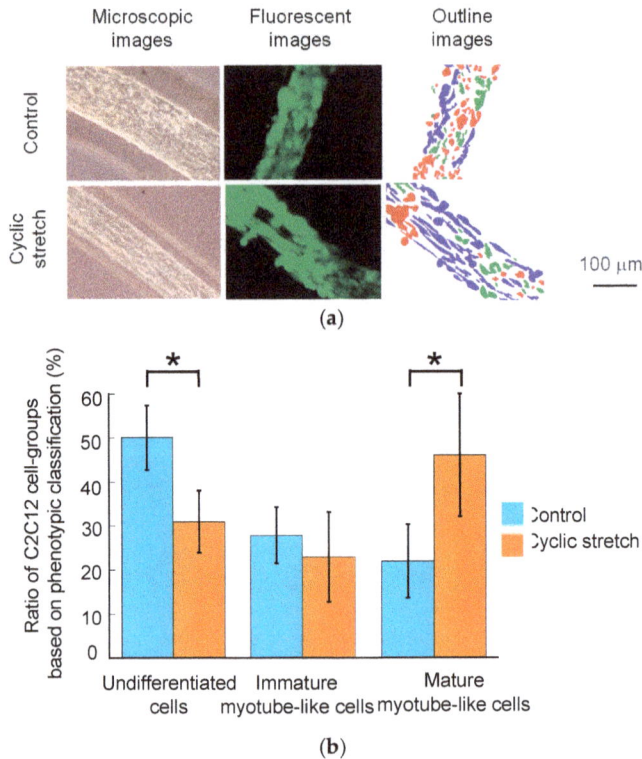

Figure 7. (a) Image-based classification of cells (blue: Mature, green: Immature, and red: Undifferentiated) and (b) the ratio of three types of C2C12 cells in the cell laden microfibers. A * indicate a significant difference ($p < 0.05$) between control and cyclic stretch groups. Scale bar: 100 μm.

In addition, for the reconstruction of the three-dimensional muscle tissue, it is important to maintain cell viability in the tissue over the culture time. In this study, the C2C12 cells were contained in the core-shell hydrogel fiber, which is suitable for three-dimensional tissue reconstruction [21]. Moreover, the hydrogel fiber structure could induce exchange of O_2, CO_2, and nutrients [32]. Therefore, our skeletal muscle reconstruction system using the cell-laden hydrogel fiber and mechanical stretching stimuli is anticipated to be applicable for in vitro tissue regeneration and clinical applications.

4. Conclusions

This study established an in vitro muscle regeneration system to use a cell-laden hydrogel fiber culture and to develop a custom-made culture device to impose the cyclic stretch stimulation on the hydrogel fiber. From the results, it was revealed that the core-shell hydrogel fiber structure could simulate "native" muscle fibrous structure to maintain the cell and muscle fiber alignment. The mechanical stretch could also promote myogenesis and maturation of muscle fibers in the cell-laden hydrogel fibers. In conclusion, our three-dimensional muscle cell culture system with mechanical stimuli could be a promising approach for tissue engineering therapy and its clinical applications.

Author Contributions: Conceptualization, H.O. and S.M.; methodology, S.B. and S.M.; validation, S.B., T.M. and S.M.; image analysis, S.B. and T.M.; investigation, S.B.; resources, S.M.; data acquisition, S.B.; writing—original draft preparation, S.M.; writing—review and editing, H.O. and S.M.; visualization, T.M. and S.M.; supervision, H.O. and S.M.; project administration, S.M.; and funding acquisition, H.O and S.M.

Funding: This research was partially supported by JSPS KAKENHI (Grant numbers: 17K01369 and 26560222) and the Translational Research Network Program from Japan Agency for Medical Research and Development (AMED).

Conflicts of Interest: H.O is a stockholder and a board member of Cellfiber Inc. which has licenses for certain cell fiber-related technologies and patents from The University of Tokyo.

References

1. Shah, R.; Knowles, J.C.; Hunt, N.P.; Lewis, M.P. Development of a novel smart scaffold for human 2 skeletal muscle regeneration. *J. Tissue Eng. Regen. Med.* **2016**, *10*, 162–171. [CrossRef] [PubMed]

2. Takahashi, H.; Okano, T. Cell Sheet-Based Tissue Engineering for Organizing Anisotropic Tissue Constructs Produced Using Microfabricated Thermoresponsive Substrates. *Adv. Healthc. Mater.* **2015**, *4*, 2388–2407. [CrossRef] [PubMed]

3. Chimenti, I.; Gaetani, R.; Barile, L.; Forte, E.; Ionta, V.; Angelini, F.; Frati, G.; Messina, E.; Giacomello, A. *Isolation and Expansion of Adult Cardiac Stem/Progenitor Cells in the Form of Cardiospheres from Human Cardiac Biopsies and Murine Hearts*; Humana Press: Totowa, NJ, USA, 2012; pp. 327–338.

4. Luo, B.; Tian, L.; Chen, N.; Ramakrishna, S.; Thakor, N.; Yang, I.H. Electrospun nanofibers facilitate better alignment, differentiation, and long-term culture in an in vitro model of the neuromuscular junction (NMJ). *Biomater. Sci.* **2018**, *6*, 3262–3272. [CrossRef] [PubMed]

5. Fee, T.; Surianarayanan, S.; Downs, C.; Zhou, Y.; Berry, J. Nanofiber Alignment Regulates NIH3T3 Cell Orientation and Cytoskeletal Gene Expression on Electrospun PCL+ Gelatin Nanofibers. *PLoS ONE* **2016**, *11*. [CrossRef] [PubMed]

6. Li, T.-S.; Cheng, K.; Lee, S.-T.; Matsushita, S.; Davis, D.; Malliaras, K.; Zhang, Y.; Matsushita, N.; Smith, R.R.; Marbán, E. Cardiospheres Recapitulate a Niche-Like Microenvironment Rich in Stemness and Cell-Matrix Interactions, Rationalizing Their Enhanced Functional Potency for Myocardial Repair. *Stem Cells* **2010**, *28*, 2088–2098. [CrossRef] [PubMed]

7. Vandenburgh, H. High-Content Drug Screening with Engineered Musculoskeletal Tissues. *Tissue Eng. Part B Rev.* **2010**, *16*, 55–64. [CrossRef] [PubMed]

8. Vandenburgh, H.; Shansky, J.; Benesch-Lee, F.; Barbata, V.; Reid, J.; Thorrez, L.; Valentini, R.; Crawford, G. Drug-screening platform based on the contractility of tissue-engineered muscle. *Muscle Nerve* **2008**, *37*, 438–447. [CrossRef]

9. Kim, W.; Kim, J.; Park, H.-S.; Jeon, J.; Kim, W.; Kim, J.; Park, H.-S.; Jeon, J.S. Development of Microfluidic Stretch System for Studying Recovery of Damaged Skeletal Muscle Cells. *Micromachines* **2018**, *9*, 671. [CrossRef]

10. Tchao, J.; Kim, J.J.; Lin, B.; Salama, G.; Lo, C.W.; Yang, L.; Tobita, K. Engineered Human Muscle Tissue from Skeletal Muscle Derived Stem Cells and Induced Pluripotent Stem Cell Derived Cardiac Cells. *Int. J. Tissue Eng.* **2013**, *2013*, 198762. [CrossRef]

11. Liu, H.; Gong, X.; Jing, X.; Ding, X.; Yao, Y.; Huang, Y.; Fan, Y. Shear stress with appropriate time-step and amplification enhances endothelial cell retention on vascular grafts. *J. Tissue Eng. Regen. Med.* **2017**, *11*, 2965–2978. [CrossRef]

12. Kaunas, R.; Nguyen, P.; Usami, S.; Chien, S. From the Cover: Cooperative effects of Rho and mechanical stretch on stress fiber organization. *Proc. Natl. Acad. Sci. USA* **2005**, *102*, 15895–15900. [CrossRef] [PubMed]

13. Wang, C.; Shan, S.; Wang, C.; Wang, J.; Li, J.; Hu, G.; Dai, K.; Li, Q.; Zhang, X. Mechanical stimulation promote the osteogenic differentiation of bone marrow stromal cells through epigenetic regulation of Sonic Hedgehog. *Exp. Cell Res.* **2017**, *352*, 346–356. [CrossRef] [PubMed]

14. Oortgiesen, D.A.W.; Yu, N.; Bronckers, A.L.J.J.; Yang, F.; Walboomers, X.F.; Jansen, J.A. A three-dimensional cell culture model to study the mechano-biological behavior in periodontal ligament regeneration. *Tissue Eng. Part C Methods* **2012**, *18*, 81–89. [CrossRef]

15. Kaji, H.; Ishibashi, T.; Nagamine, K.; Kanzaki, M.; Nishizawa, M. Electrically induced contraction of C2C12 myotubes cultured on a porous membrane-based substrate with muscle tissue-like stiffness. *Biomaterials* **2010**, *31*, 6981–6986. [CrossRef]

16. Grossi, A.; Lametsch, R.; Karlsson, A.H.; Lawson, M.A. Mechanical stimuli on C2C12 myoblasts affect myoblast differentiation, focal adhesion kinase phosphorylation and galectin-1 expression: A proteomic approach. *Cell Biol. Int.* **2011**, *35*, 579–586. [CrossRef] [PubMed]

17. Kasper, A.M.; Turner, D.C.; Martin, N.R.W.; Sharples, A.P. Mimicking exercise in three-dimensional bioengineered skeletal muscle to investigate cellular and molecular mechanisms of physiological adaptation. *J. Cell. Physiol.* **2018**, *233*, 1985–1998. [CrossRef] [PubMed]

18. Hornberger, T.A.; Armstrong, D.D.; Koh, T.J.; Burkholder, T.J.; Esser, K.A. Intracellular signaling specificity in response to uniaxial vs. multiaxial stretch: Implications for mechanotransduction. *Am. J. Physiol. Cell Physiol.* **2005**, *288*, C185–C194. [CrossRef] [PubMed]

19. Asano, S.; Ito, S.; Morosawa, M.; Furuya, K.; Naruse, K.; Sokabe, M.; Yamaguchi, E.; Hasegawa, Y. Cyclic stretch enhances reorientation and differentiation of 3-D culture model of human airway smooth muscle. *Biochem. Biophys. Rep.* **2018**, *16*, 32–38. [CrossRef] [PubMed]

20. Nguyen, T.D.; Liang, R.; Woo, S.L.-Y.; Burton, S.D.; Wu, C.; Almarza, A.; Sacks, M.S.; Abramowitch, S. Effects of cell seeding and cyclic stretch on the fiber remodeling in an extracellular matrix-derived bioscaffold. *Tissue Eng. Part A* **2009**, *15*, 957–963. [CrossRef] [PubMed]

21. Onoe, H.; Okitsu, T.; Itou, A.; Kato-Negishi, M.; Gojo, R.; Kiriya, D.; Sato, K.; Miura, S.; Iwanaga, S.; Kuribayashi-Shigetomi, K.; et al. Metre-long cell-laden microfibres exhibit tissue morphologies and functions. *Nat. Mater.* **2013**, *12*, 584–590. [CrossRef] [PubMed]

22. Li, R.H.; Altreuter, D.H.; Gentile, F.T. Transport characterization of hydrogel matrices for cell encapsulation. *Biotechnol. Bioeng.* **1996**, *50*, 365–373. [CrossRef]

23. Park, H.; Bhalla, R.; Saigal, R.; Radisic, M.; Watson, N.; Langer, R.; Vunjak-Novakovic, G. Effects of electrical stimulation in C2C12 muscle constructs. *J. Tissue Eng. Regen. Med.* **2008**, *2*, 279–287. [CrossRef] [PubMed]

24. Yamasaki, K.; Hayashi, H.; Nishiyama, K.; Kobayashi, H.; Uto, S.; Kondo, H.; Hashimoto, S.; Fujisato, T. Control of myotube contraction using electrical pulse stimulation for bio-actuator. *J. Artif. Organs* **2009**, *12*, 131–137. [CrossRef] [PubMed]

25. Fujita, H.; Endo, A.; Shimizu, K.; Nagamori, E. Evaluation of serum-free differentiation conditions for C2C12 myoblast cells assessed as to active tension generation capability. *Biotechnol. Bioeng.* **2010**, *107*, 894–901. [CrossRef] [PubMed]

26. Chang, Y.-J.; Chen, Y.-J.; Huang, C.-W.; Fan, S.-C.; Huang, B.-M.; Chang, W.-T.; Tsai, Y.-S.; Su, F.-C.; Wu, C.-C. Cyclic Stretch Facilitates Myogenesis in C2C12 Myoblasts and Rescues Thiazolidinedione-Inhibited Myotube Formation. *Front. Bioeng. Biotechnol.* **2016**, *4*, 27. [CrossRef] [PubMed]

27. Evans, D.J.; Britland, S.; Wigmore, P.M. Differential response of fetal and neonatal myoblasts to topographical guidance cues in vitro. *Dev. Genes Evol.* **1999**, *209*, 438–442. [CrossRef] [PubMed]

28. Lam, M.T.; Sim, S.; Zhu, X.; Takayama, S. The effect of continuous wavy micropatterns on silicone substrates on the alignment of skeletal muscle myoblasts and myotubes. *Biomaterials* **2006**, *27*, 4340–4347. [CrossRef] [PubMed]

29. Guex, A.G.; Birrer, D.L.; Fortunato, G.; Tevaearai, H.T.; Giraud, M.-N. Anisotropically oriented electrospun matrices with an imprinted periodic micropattern: A new scaffold for engineered muscle constructs. *Biomed. Mater.* **2013**, *8*, 021001. [CrossRef] [PubMed]

30. Lam, M.T.; Huang, Y.-C.; Birla, R.K.; Takayama, S. Microfeature guided skeletal muscle tissue engineering for highly organized 3-dimensional free-standing constructs. *Biomaterials* **2009**, *30*, 1150–1155. [CrossRef] [PubMed]

31. Tanaka, T.; Hattori-Aramaki, N.; Sunohara, A.; Okabe, K.; Sakamoto, Y.; Ochiai, H.; Hayashi, R.; Kishi, K. Alignment of Skeletal Muscle Cells Cultured in Collagen Gel by Mechanical and Electrical Stimulation. *Int. J. Tissue Eng.* **2014**, *2014*, 621529. [CrossRef]

32. Ikeda, K.; Nagata, S.; Okitsu, T.; Takeuchi, S. Cell fiber-based three-dimensional culture system for highly efficient expansion of human induced pluripotent stem cells. *Sci. Rep.* **2017**, *7*, 2850. [CrossRef] [PubMed]

![micromachines logo] *micromachines*

MDPI

Article

Temporal Observation of Adipocyte Microfiber Using Anchoring Device

Akiyo Yokomizo [1], Yuya Morimoto [1,2], Keigo Nishimura [1,3] and Shoji Takeuchi [1,2,3,4,*]

[1]   Center for International Research on Integrative Biomedical Systems (CIBiS), Institute of Industrial Science (IIS), The University of Tokyo, 4-6-1 Komaba, Meguro-ku, Tokyo 153-8505, Japan; yokomizo@iis.u-tokyo.ac.jp (A.Y.); y-morimo@hybrid.t.u-tokyo.ac.jp (Y.M.); nishimura@hybrid.t.u-tokyo.ac.jp (K.N.)

[2]   Department of Mechano-Informatics, Graduate School of Information Science and Technology, The University of Tokyo, 7-3-1 Hongo, Bunkyo-ku, Tokyo 113-8656, Japan

[3]   Department of Life Sciences, Graduate School of Arts and Sciences, The University of Tokyo, 7-3-1 Hongo, Bunkyo-ku, Tokyo 113-8656, Japan

[4]   International Research Center for Neurointelligence (WPI-IRCN), The University of Tokyo Institutes for Advanced Study (UTIAS), The University of Tokyo, 7-3-1 Hongo, Bunkyo-ku, Tokyo 113-8656, Japan

[*]   Correspondence: takeuchi@hybrid.t.u-tokyo.ac.jp; Tel.: +81-3-5841-6488; Fax: +81-3-5841-6199

Received: 11 May 2019; Accepted: 28 May 2019; Published: 29 May 2019

Abstract: In this paper, we propose an anchoring device with pillars to immobilize an adipocyte microfiber that has a fiber-shaped adipocyte tissue covered by an alginate gel shell. Because the device enabled the immobilization of the microfiber in a culture dish even after its transportation and the exchange of the culture medium, we can easily track the specific positions of the microfiber for a long period. Owing to the characteristics of the anchoring device, we successfully performed temporal observations of the microfiber on the device for a month to investigate the function and morphology of three-dimensional cultured adipocytes. Furthermore, to demonstrate the applicability of the anchoring device to drug testing, we evaluated the lipolysis of the microfiber's adipocytes by applying reagents with an anti-obesity effect. Therefore, we believe that the anchoring device with the microfiber will be a useful tool for temporal biochemical analyses.

Keywords: microfluidics; biofabrication; adipose tissue; lipolysis

1. Introduction

Core-shell cell microfibers, a fiber-shaped cellular-tissue covered by a shell of alginate gel, have become attractive in various applications, such as tissue engineering, cell therapy, and drug testing [1], because the three-dimensional (3D) culture of cells are performed at the core, and the alginate gel shell protects the cells from physical stimuli while allowing the diffusion of nutrition and oxygen [2]. Hence, core–shell cell microfibers have been widely used to construct various types of tissues including connective [2–5], neural [6], stem cell [7–10], smooth muscle [11], and adipose tissues [12]. Particularly, the culture of adipocytes in microfiber is a promising approach in the construction of adipose tissue, as the microfiber maintains the 3D culture of adipocytes by covering with the shell of the alginate gel for a long period. Therefore, core–shell adipocyte microfibers achieved the formation of large lipid droplets comparable to living adipose tissues. The adipocyte microfiber has advantages in culture dimension, handleability, and high-throughput production [12], compared to the recent methods for the adipocyte tissue formation with synthetic scaffolds and structures of extracellular matrix [13]. Moreover, conventional two-dimensional (2D) culture methods induce the detachment of adipocytes from a culture dish because of the increase of the adipocyte buoyancy during lipid accumulation [13,14]. Meanwhile, although the 2D culture enables the studies of lipid metabolism [15–17], the adipocyte microfibers cannot investigate the time course of changes in the function and morphology of the adipocytes because the microfibers move freely in a culture medium during their transportation and the exchange of medium.

In this paper, we propose a donut-shaped anchoring device with pillars to tangle an adipocyte microfiber (Figure 1). By placing the microfiber at a hollow section in the device and immobilizing both ends of the microfiber at the pillars, the microfiber can be observed clearly for a long period without damaging the cells in the microfiber. In addition, by referring to the position guides of the anchoring device, tracking a specific site in the microfiber is possible even after placing it out of the field of view of the microscope. Here, to evaluate the characteristics of the anchoring device, we set the adipocyte microfibers on the device and investigate the effect on the maturation of the adipocytes. Furthermore, by long periods of clear observations, we demonstrate that the anchoring device facilitates in the continuous observation of cellular morphology in the microfiber and evaluation of fatty acid release from the adipocyte microfiber by applying reagents, as an example of a drug testing application.

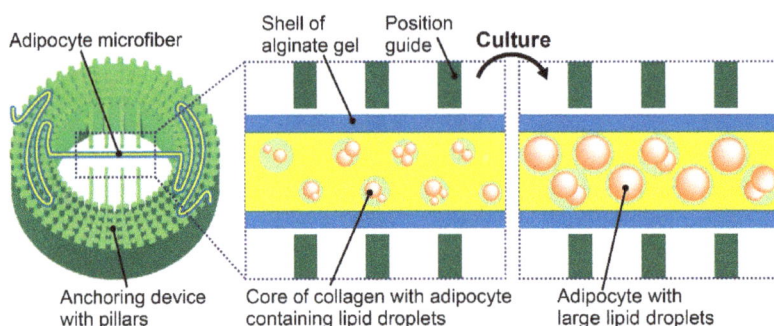

Figure 1. Conceptual illustration for immobilization of an adipocyte microfiber using the proposed anchoring device. By tangling the microfiber with pillars on the device manually, the microfiber becomes observable at the hollow section of the device for a long term without damaging the adipocytes.

2. Materials and Methods

2.1. Cell Preparation

3T3-L1 cells (mouse adipocytes, JCRB Cell Bank, Osaka, Japan, Cell No. JCRB9014) were seeded and maintained according to the manufacturer's instructions, using a growth medium that was Dulbecco's modified eagle medium low glucose (DMEM LG, 041-29775, FUJIFILM Wako Pure Chemical Corp., Osaka, Japan) containing 10% (v:v) fetal bovine serum (FBS (Chile Origin, USDA approved), FB-1365/500, Biosera, Nuaille, France), and 1% (v:v) penicillin/streptomycin (P/S, Sigma-Aldrich, St. Louis, MO, USA) at 37 °C in a 5% CO_2 atmosphere. Passages were performed before the confluence of the cells.

2.2. Fabrication of the Anchoring Device

The anchoring device for the immobilization of an adipocyte microfiber is 17 mm in outer diameter, 10 mm in inner diameter, and 6 mm in height (Figure 2); it is designed to be placed within a 35-mm culture dish and not to be floated in 2 mL of culture medium. Furthermore, 50 patterns of 5 pillars (500 μm in diameter, 1 mm in height) with 300-μm intervals were placed on the top surface of the device. The intervals were almost the same as the diameter of the microfiber. Bars of 0.3 mm width were arranged at the center of the device and used as a position guide for the identification of the observation area. The anchoring device was fabricated using a 3D printer (Perfactory 4 mini, Envision TEC, Dearborn, MI, USA). The fabricated device was coated with a 2-μm parylene layer using a chemical vapor deposition machine (Parylene Deposition System 2010, Specialty Coating Systems, Inc., Indianapolis, IN, USA) to improve the cell compatibility of the device [18,19].

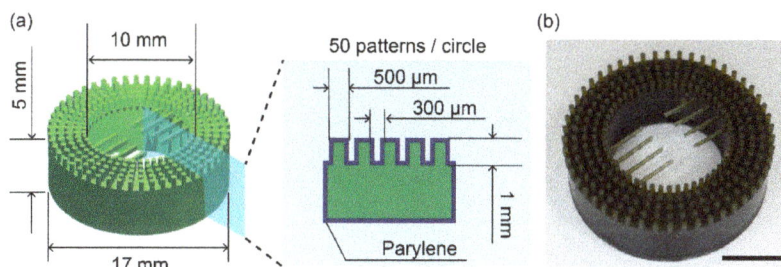

Figure 2. Design of the anchoring device with pillars: (**a**) Schematic illustration of the anchoring device with dimensions; (**b**) Image of the fabricated anchoring device. Scale bar is 5 mm.

2.3. Formation of Adipocyte Microfibers

We fabricated adipocyte-laden hydrogel microfibers using a previously proposed method with a triple coaxial microfluidic device [2]. In the formation of the microfibers, a core solution, shell solution, and sheath solution were infused into the innermost channel, intermediate channel, and outermost channel of the microfluidic device, respectively. The flow rate of the core solution, i.e., a collagen solution (I-AC 50, KOKEN Co., Ltd., Tokyo, Japan) with 3T3-L1 cells at 1.0×10^8 cells/mL, was 50 to 150 µL/min and that of the shell solution, i.e., a 1.5 wt% sodium alginate solution (194-13321, FUJIFILM Wako Pure Chemical Corp., Osaka, Japan), was 300 µL/min. The flow rate of the sheath solution, i.e., a 100-mM calcium chloride solution (191-01665, FUJIFILM Wako Pure Chemical Corp., Osaka, Japan) for gelling sodium alginate, was 3600 µL/min. After infusing the solutions into the microfluidic device, an adipocyte-laden hydrogel microfiber was formed; the fiber comprises an inner fiber-shaped adipocyte-laden collagen gel covered by an alginate gel layer. Subsequently, we placed the fabricated microfiber in the growth medium. After 2 days of culture, the growth medium was replaced with a differentiation medium (Preadipocyte Growth Medium-2 BulletKit™, PT-8002, Lonza, Basel, Switzerland). After 2 days of the culture with the differentiation medium, the medium was changed to a maturation medium that was DMEM high glucose (D5796, Sigma-Aldrich, St. Louis, MO, USA) containing 10% (v:v) FBS, 1% (v:v) P/S, and 5 µg/mL insulin (10516, Sigma-Aldrich, St. Louis, MO, USA). The maturation medium was replaced with a fresh medium every three days. Finally, the adipocytes were in contact with each other in the microfiber, thus resulting in the formation of adipocyte microfibers composed of a fiber-shaped adipocyte tissue covered by an alginate gel layer.

2.4. Microfiber Immobilization

For the immobilization of the adipocyte microfiber with the anchoring device, we first cut the microfiber to a length of ~5 cm with a pair of scissors such that the microfiber is mountable to the device (Figure 3a). To increase the strength of the alginate gel layer of the microfiber, we transferred it to a new dish containing a 1 mL aliquot of a 100-mM calcium chloride solution and 10-mL maturation medium (Figure 3b). Next, as a pretreatment of immobilization, we wetted a 35-mm dish containing the device with the culture medium to prevent the microfiber from adhering to the bottom of the dish; we also wetted the pillars on the device (Figure 3c). In addition, we used a static electricity removal gun and stopped the fan of the clean bench to prevent the microfiber from adhering to unintended positions. Subsequently, the microfiber was manually tangled with pillars using a pair of tweezers (Figure 3d). In this state, the surface tension of the culture medium between the pillars pushed the microfiber to the top surface of the device. Finally, to facilitate an observation with a microscope, we turned over the device on the dish bottom (Figure 3e). Then, the culture medium was injected gently from the outside of the device for the culture.

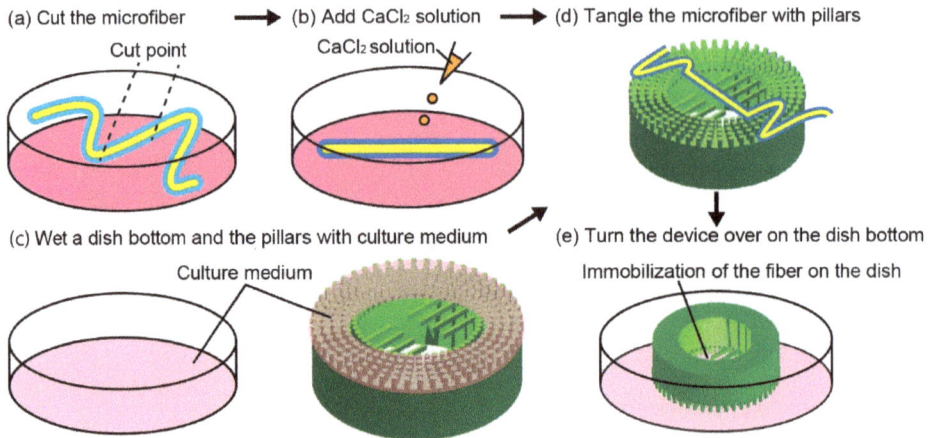

Figure 3. Process flow of immobilization of an adipocyte microfiber with an anchoring device: (**a**) Cut the microfiber with scissors; (**b**) Add CaCl₂ solution to increase the strength of the alginate gel layer; (**c**) Wet a dish bottom and the pillars with a culture medium; (**d**) Tangle the microfiber with pillars using a pair of tweezers; (**e**) Turn the device over on the dish bottom for a clear observation.

To verify the immobility of the microfiber on the anchoring device during culture, we observed the movements of a fluorescent alginate hydrogel microfiber in the presence of disturbances. The fluorescent alginate hydrogel microfiber was fabricated by infusing a 1.5 wt% sodium alginate solution containing 1 µL/mL red fluorescent beads (F8810, Life Technologies Corp., Carlsbad, CA, USA) as a core solution and a 1.5 wt% sodium alginate solution with 1 µL/mL green fluorescent beads (F8811, Life Technology, Inc.) as a shell solution into the microfluidic device. When we applied a flow to the device as a disturbance, we connected syringes to the dish using septums and Teflon tubes (Figure 4). First, a 5-mm-thick silicone rubber sponge (5-3030-04, AS ONE, Osaka, Japan) was punched out with a trepan for biopsy, and 2 pieces of 4-mm-diameter rubber sponges with a 1-mm-diameter hole at the center were prepared. Subsequently, we pressed them into 2.5-mm diameter holes that were made on the cover of a 35-mm culture dish. Finally, we inserted the tubes into the septums and connected them to the syringe through silicone tubes. Then, we flowed sterilized water at 0.01 to 10 mL/min for 60 s and measured the distances between the guides of the device and the microfiber using a microscope (IX71N, Olympus Corp., Tokyo, Japan) and imaging software (Image J, 1.46r software package, National Institutes of Health (NIH), Bethesda, MD, USA). The distances were labeled as d_1, d_2, d_3, and d_4 in order from the left side of the images. In the case of applying rotations to the device, the dish with the device was placed on a turntable (NA-301, Nissinrika, Tokyo, Japan) and rotated at 30 to 120 rpm for 30 s. After the rotation, we measured the distances between the guides and microfibers using the same method as the measurement under an applied flow.

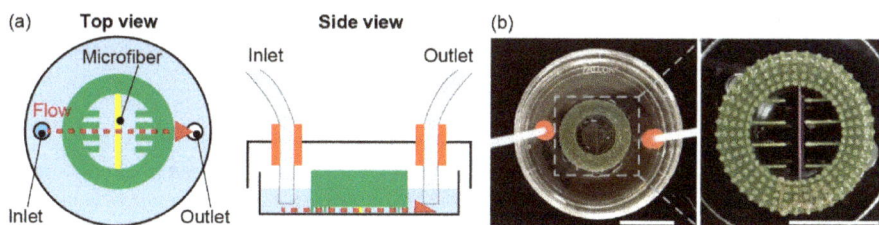

Figure 4. Experimental setup to evaluate immobilization of the microfiber in the anchoring device: (**a**) Conceptual illustration of applying flow to the microfiber in the device; (**b**) Image of the experimental setup. Scale bars are 10 mm.

2.5. Morphology Evaluation of the Microfiber

To evaluate the dimensions of the adipocyte-laden hydrogel microfibers according to the flow rates of the infused flows in the microfluidic device, we prepared the microfibers with different flow rates of core solution (50 µL/min, 100 µL/min, 150 µL/min), and measured the diameters of the shell and core of the microfibers using a microscope (IX71N, Olympus Corp., Tokyo, Japan) and an imaging software (cellSens, Olympus Corp, Tokyo, Japan). Moreover, we measured the sizes of lipid droplets in the adipocytes of the cultured adipocyte microfibers using bright-field images or fluorescent images in the case of staining the lipid droplets with BODIPY 493/503 (D3922, Thermo Fisher Scientific, Waltham, MA, USA). In the measurement, we defined the diameter of the lipid droplets as the average value of the orthogonal diameters measured on an imaging software (Image J 1.46r software package, National Institutes of Health (NIH), Bethesda, MD, USA). To compare the lipid droplet size between the adipocyte microfibers and the 2D cultured adipocytes, we seeded the adipocytes to a dish at 18,000 cells/cm^2 and changed the growth medium to the differentiation medium after confluence. After the adipocytes were cultured on a dish, we measured the diameters of the lipid droplets in the adipocytes using the same method as that for the adipocyte fibers.

To investigate the effect of the anchoring device to the culture of the adipocyte microfibers, we measured the size of the lipid droplets in the microfibers cultured on the device using the method above. In addition, we set an adipocyte microfiber on the anchoring device to observe the edge of an adipocyte tissue and verified the morphological changes in lipid droplets of the tissue. In the experiment, we used the maturation medium containing oleic acid-BSA complex (O3008, Sigma-Aldrich, St. Louis, MO, USA) at 500 µM. Five minutes after culture in the culture medium, we started the observation for the lipid droplets using an all-in-one microscope (BZ 9000, KEYENCE Corp., Osaka, Japan) and captured their images hourly for 24 h.

2.6. Evaluation of Lipolysis in Adipocyte Microfiber

To demonstrate the applicability of the adipocyte microfiber immobilized with the anchoring device to drug testing, we evaluated the lipolysis of the adipocytes in the microfiber. In the experiment, we first cultured the microfiber with 2 mL of a DMEM high glucose solution to prevent the effects of serum and insulin and 1 µL of 1 mg/mL BODIPY to dye the lipid droplets. After incubation for 3 h or more, we changed the culture medium to DMEM low glucose solution containing 2% (v:v) BSA (A7906, Sigma-Aldrich) and 5% (v:v) FBS. 5 min after the medium change, we started the observation of the microfiber using the all-in-one microscope and captured fluorescence images hourly for 6 h. Subsequently, we changed the culture medium to DMEM low glucose solution containing 2% (v:v) BSA, 5% (v:v) FBS, and 1 mM isoproterenol (I5627, Sigma-Aldrich, St. Louis, MO, USA). After 5 min of culture, we recorded fluorescence images hourly for 6 h. To get multiple data, we repeated the experiment after 24 h and more from the previous experiment. After repeating experiments three times, we extracted green-channel images from the captured full-color images and measured the fluorescence intensity of the microfiber using Image J.

3. Results and Discussion

3.1. Characterization of the Anchoring Device

To verify the characteristics of the anchoring device for the immobilization of the adipocyte microfiber, we applied flows to the device with the fluorescent hydrogel fiber (Figure 5a). Consequently, we confirmed that the moving distance of the hydrogel microfiber induced by the flows was within 50 μm under a flow rate lower than 1 mL/min (Figure 5b). Meanwhile, when the flow rate was 10 mL/min, the hydrogel microfiber was immobilized on the device. However, the device moved, thus causing the microfiber to be outside the microscope's field of view. In the case of applying rotations to the device, the moving distance was within 100 μm even when the rotation speed was 120 rpm (Figure 5c). These results indicate that our anchoring device can maintain the position of the microfiber even under disturbances, such as transportation of the culture dish and exchanges in the culture medium. Therefore, the anchoring device allows us to trace the specific position of the adipocyte microfiber during culture. Although immobilization has been achieved by reeling and weaving the microfibers and sucking both ends of the microfibers [2–4,10,20–25], both methods are not suitable for investigating the time course of cellular changes in the microfibers. The reeling and weaving cause difficulties in maintaining a clear observation because the microfibers may overlap each other. While sucking can immobilize the microfibers to avoid overlapping, it is not suitable for the investigation during culture because a culture medium is also sucked during the immobilization, resulting in dynamic changes in culture conditions. Thus, we believe that the anchoring device is a useful tool for the investigation of the microfibers by temporal observation.

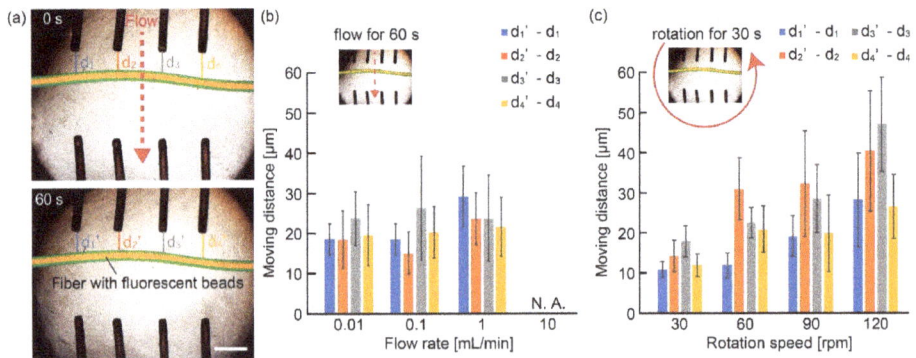

Figure 5. Evaluation of immobilization of a microfiber with fluorescent beads with the anchoring device: (**a**) Images of the microfiber before and after applying a 1 mL/min flow in the device; (**b,c**) Moving distance of the microfiber in the anchoring device when we applied (**b**) a flow and (**c**) a rotation to the device (N = 5, mean ± s. d.). Scale bar is 1 mm.

3.2. Characterization of Adipocyte Microfiber

To investigate the appropriate fabrication conditions for the formation of the adipocyte microfiber, we prepared the adipocyte-laden hydrogel microfibers under different flow rates of the core solution. Although the diameter of the whole fiber (the diameter of the alginate gel shell) was not changed significantly with the increase in the flow rate of the core solution, the diameter of the adipocyte-laden collagen core was increased (Figure 6a,b). This result indicates that it is possible to control the thicknesses of the shell and core without changing the diameter of the whole fiber by controlling the flow rate of the core solution, to induce changes in the culture condition for adipocytes under maintenance of mountable microfiber diameter (~ 300 μm) to the anchoring device. As a result of culturing for the maturation of adipocytes, we recognized that the adipocyte microfiber fabricated at 100 μL/min of the core solution exhibited larger lipid droplets than those fabricated at 50 and

150 μL/min of the core solution (Figure 6c). Therefore, in this paper, we decided to use the adipocyte microfiber prepared with 100 μL/min of the core solution in the following experiments.

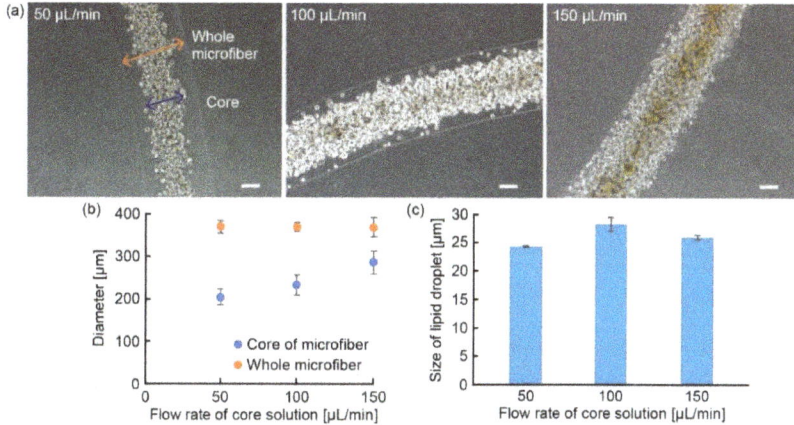

Figure 6. Relationship between flow rate of core solution and size of lipid droplets in adipocytes: (a) Images of adipocyte-laden hydrogel microfibers (the flow rate of core solution was 50 μL/min, 100 μL/min, 150 μL/min); (b) Relationship between diameters of the microfibers and flow rates of core solution under 300 μL/min flow of shell solution (N ≥ 5, mean ± s. e. m.); (c) Size of lipid droplets after 21 days of culture depending on the flow rate of the core solution (N = 3, mean ± s. e. m.). Scale bars are 100 μm.

To evaluate the effects of the culture using microfiber for the adipocytes, we compared the morphology of the adipocytes cultured in the microfiber with those cultured in a 2D culture dish. While the adipocytes in 2D culture became sparse, the adipocytes in the microfiber became dense adipose tissue as in vivo (Figure 7a). The average lipid droplet size of the microfiber was 28 ± 13 μm, which was approximately twice larger than that of the 2D culture, 13 ± 7 μm (mean ± s. d.) (Figure 7b). In addition, the peak of the size distribution of the lipid droplets in 2D culture was as small as 10 μm, while the size of lipid droplets in the microfiber was distributed widely from 10 μm to 50 μm (Figure 7c). The formation of large lipid droplets in the adipocytes is important to mimic the state of obesity because the responses of adipocytes to biochemicals change depending on the size of the lipid droplets [26,27]. Therefore, these results indicate that the adipocyte microfiber provides an adipocyte tissue which is more useful in the investigation of biochemical reactions under obesity than 2D cultured adipocytes.

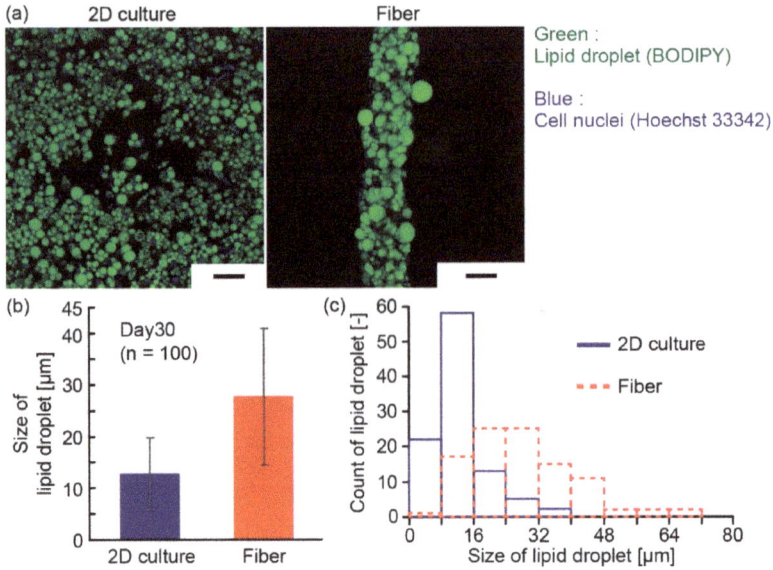

Figure 7. Comparison of 2D-cultured adipocytes in a dish and adipocytes in a microfiber: (**a**) Fluorescence images of lipid droplets in 2D culture and microfiber culture; (**b**) Average size (n = 100, mean ± s. d.) and (**c**) size distribution of lipid droplets on day 30 in both culture conditions. Scale bars are 100 μm.

Furthermore, we investigated the culture characteristics of the adipocyte microfiber on the anchoring device. By tangling the microfiber with pillars on the device, we achieved the immobilization of the adipocyte microfiber and clear observation of the adipocytes (Figure 8a). Owing to the features of the device, we successfully evaluated the time-dependent change of the size of the lipid droplets for a month (Figure 8b). In the comparison of the lipid droplet sizes when cultured with or without the device, no significant difference was indicated ($p > 0.2$, Student's *t*-test) (Figure 8c). This result indicates that the anchoring device does not affect the maturation of the adipocytes in the microfiber. Moreover, we continuously observed the edge of an adipocyte tissue in an adipocyte microfiber. Consequently, we confirmed the migration of lipid droplets, changes in their size, and fusion of the droplets in the adipocyte microfiber (Figure 9). This result indicates that the anchoring device allows us to observe the behaviors of single adipocytes in the microfiber, leading to studies of lipid metabolism based on the morphology of the lipid droplets.

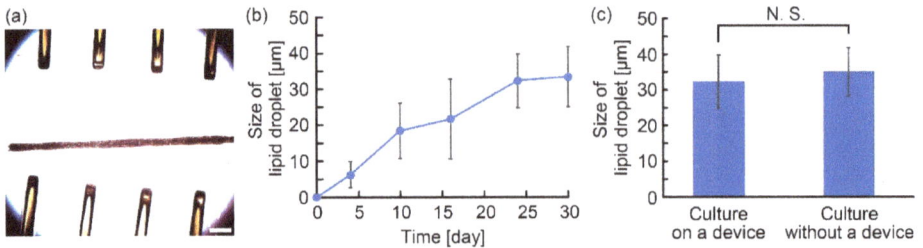

Figure 8. Observation of an adipocyte microfiber cultured on the anchoring device: (**a**) Images of an adipocyte microfiber on day 28; (**b**) Size of lipid droplets varying with time (n ≥ 10, mean ± s. d.); (**c**) Comparison of lipid droplet size in adipocyte microfibers cultured with or without the anchoring device for 24 days (n = 20, mean ± s. d.). Scale bar is 500 μm.

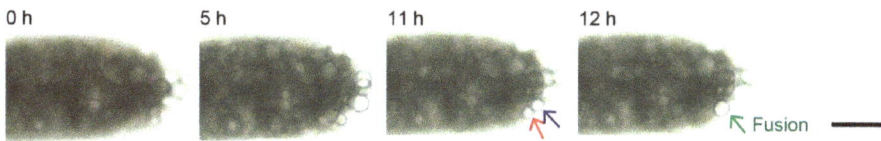

Figure 9. Time-lapse images of an adipocyte microfiber on the anchoring device. A fusion (green arrow) of lipid droplets (red and blue arrows) was observed. Scale bar is 100 μm.

3.3. Lipolysis of Adipocyte Microfiber

To demonstrate the applicability of the adipocyte microfiber immobilized with the anchoring device to drug testing, we examined the lipolysis of the adipocyte microfiber using isoproterenol, a reagent with anti-obesity effect. When we added isoproterenol to the microfiber stained with BODIPY for the visualization of lipids on the device, the fluorescence intensity of the microfiber decreased as time progressed, in contrast to the fluorescence intensity of the microfiber without the addition of isoproterenol (Figure 10). As the exposure times were the same in both experiments, the decrease in fluorescence intensity was not caused by the quenching of BODIPY with laser irradiation. Therefore, the results indicate that the decrease was caused by the release of fatty acids from the adipocytes, demonstrating that lipolysis by isoproterenol was achieved in the microfiber. Hence, the anchoring device enabled the observation of lipolysis by the immobilization of the microfiber. Therefore, the feature of the device enabling the temporal observation of the specific position in the microfiber is useful in lipid metabolism research or early-stage drug screening.

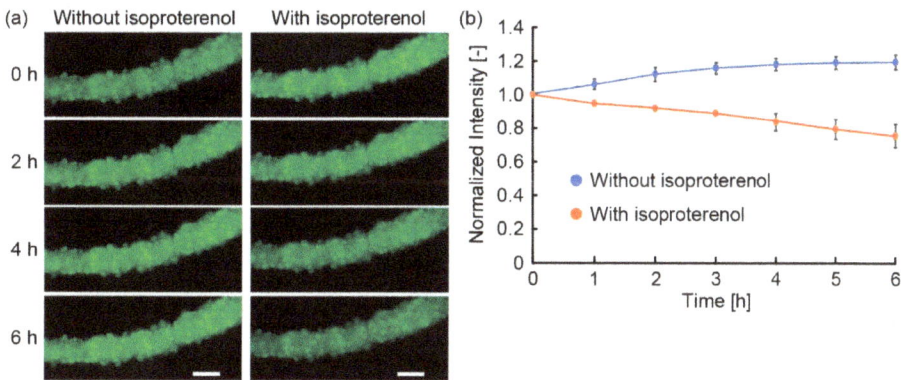

Figure 10. Changes in fluorescence intensity of lipid droplets stained with BODIPY when applying isoproterenol to the adipocyte microfiber on the anchoring device: (**a**) Time-lapse fluorescent images of the microfiber on the anchoring device; (**b**) Variation with time of fluorescence intensity of the adipocyte microfiber in specific observation area under repeating the experiment three times (mean ± s. d.). The fluorescent intensity is normalized with that at 0 h. Scale bars are 100 μm.

4. Conclusions

In this study, we developed an anchoring device with pillars for the temporal observation of an adipocyte microfiber. The advantages of the anchoring devices are as follows: (i) temporal observation of the specific position in the microfiber even when applying disturbances; (ii) maturation of adipocytes in the microfiber on the anchoring device comparing to that without the device; (iii) clear observation of lipid droplets and cell morphology in the microfiber. Based on these advantages, we show that the adipocyte microfiber immobilized on the anchoring device could be used for drug testing by enabling the study of lipolysis under the addition of isoproterenol. Although we focused on the adipocyte microfiber herein, the anchoring device enabled the temporal observation of various types of cell

Micromachines **2019**, *10*, 358

microfibers. Therefore, we believe that the anchoring device will be a useful tool for studies in various fields, such as biochemistry and drug testing.

Author Contributions: A.Y., Y.M., and S.T. conceived and designed the experiments; A.Y. and K.N. performed the experiments and analyzed the data; A.Y., Y.M., and S.T. wrote the paper. All authors discussed the results and contributed to the manuscript.

Funding: This work was partially supported by JSPS Core-to-Core Program, JST-Mirai Program Grant Number JPMJMI18CE, and JSPS KAKENHI Grant Number 16H06329.

Acknowledgments: The authors thank Akane Itou for her technical support and Minghao Nie for his valuable comments.

Conflicts of Interest: S.T. is an inventor on intellectual property rights related to the cell fiber technology, and stockholders of Cellfiber Inc., a start-up company based on the cell fiber technology.

References

1. Morimoto, Y.; Hsiao, A.Y.; Takeuchi, S. Point-, line-, and plane-shaped cellular constructs for 3D tissue assembly. *Adv. Drug Deliver. Rev.* **2015**, *95*, 29–39. [CrossRef] [PubMed]
2. Onoe, H.; Okitsu, T.; Itou, A.; Kato-Negishi, M.; Gojo, R.; Kiriya, D.; Sato, K.; Miura, S.; Iwanaga, S.; Kuribayashi-Shigetomi, K.; et al. Metre-long cell-laden microfibres exhibit tissue morphologies and functions. *Nat. Mater.* **2013**, *12*, 584–590. [CrossRef] [PubMed]
3. Mistry, P.; Aied, A.; Alexander, M.; Shakesheff, K.; Bennett, A.; Yang, J. Bioprinting Using Mechanically Robust Core–shell Cell-Laden Hydrogel Strands. *Macromol. Biosci.* **2017**, *17*, 1600472. [CrossRef] [PubMed]
4. Liu, W.J.; Zhong, Z.; Hu, N.; Zhou, Y.X.; Maggio, L.; Miri, A.K.; Fragasso, A.; Jin, X.Y.; Khademhosseini, A.; Zhang, Y.S. Coaxial extrusion bioprinting of 3D microfibrous constructs with cell-favorable gelatin methacryloyl microenvironments. *Biofabrication* **2018**, *10*, 024102. [CrossRef] [PubMed]
5. Morimoto, Y.; Kiyosawa, M.; Takeuchi, S. Three-dimensional printed microfluidic modules for design changeable coaxial microfluidic devices. *Sens. Actuators B* **2018**, *274*, 491–500. [CrossRef]
6. Onoe, H.; Kato-Negishi, M.; Itou, A.; Takeuchi, S. Differentiation Induction of Mouse Neural Stem Cells in Hydrogel Tubular Microenvironments with Controlled Tube Dimensions. *Adv. Healthc. Mater.* **2016**, *5*, 1104–1111. [CrossRef]
7. Tian, C.; Zhang, X.; Zhao, G. Vitrification of stem cell-laden core-shell microfibers with unusually low concentrations of cryoprotective agents. *Biomater. Sci.* **2019**, *7*, 889–900. [CrossRef]
8. Ikeda, K.; Nagata, S.; Okitsu, T.; Takeuchi, S. Cell fiber-based three-dimensional culture system for highly efficient expansion of human induced pluripotent stem cells. *Sci. Rep.* **2017**, *7*, 2850. [CrossRef]
9. Perez, R.A.; Kim, M.; Kim, T.H.; Kim, J.H.; Lee, J.H.; Park, J.H.; Knowles, J.C.; Kim, H.W. Utilizing Core-Shell Fibrous Collagen-Alginate Hydrogel Cell Delivery System for Bone Tissue Engineering. *Tissue Eng. Part A* **2014**, *20*, 103–114. [CrossRef]
10. Dai, X.L.; Liu, L.B.; Ouyang, J.; Li, X.D.; Zhang, X.Z.; Lan, Q.; Xu, T. Coaxial 3D bioprinting of self-assembled multicellular heterogeneous tumor fibers. *Sci. Rep.* **2017**, *7*, 1457. [CrossRef]
11. Hsiao, A.Y.; Okitsu, T.; Onoe, H.; Kiyosawa, M.; Teramae, H.; Iwanaga, S.; Kazama, T.; Matsumoto, T.; Takeuchi, S. Smooth Muscle-Like Tissue Constructs with Circumferentially Oriented Cells Formed by the Cell Fiber Technology. *PLoS ONE* **2015**, *10*, e0119010. [CrossRef] [PubMed]
12. Hsiao, A.Y.; Okitsu, T.; Teramae, H.; Takeuchi, S. 3D Tissue Formation of Unilocular Adipocytes in Hydrogel Microfibers. *Adv. Healthc. Mater.* **2016**, *5*, 548–556. [CrossRef] [PubMed]
13. Pope, B.D.; Warren, C.R.; Parker, K.K.; Cowan, C.A. Microenvironmental control of adipocyte fate and function. *Trends Cell Biol.* **2016**, *26*, 745–755. [CrossRef] [PubMed]
14. Zhang, Z.Z.; Kumar, S.; Barnett, A.H.; Eggo, M.C. Ceiling culture of mature human adipocytes: use in studies of adipocyte functions. *J. Endocrinol.* **2000**, *164*, 119–128. [CrossRef] [PubMed]
15. Paar, M.; Jüngst, C.; Steiner, N.A.; Magnes, C.; Sinner, F.; Kolb, D.; Lass, A.; Zimmermann, R.; Zumbusch, A.; Kohlwein, S.D.; et al. Remodeling of Lipid Droplets during Lipolysis and Growth in Adipocytes. *J. Biol. Chem.* **2012**, *287*, 11164–11173. [CrossRef] [PubMed]
16. Miller, C.N.; Yang, J.-Y.; England, E.; Yin, A.; Baile, C.A.; Rayalam, S. Isoproterenol Increases Uncoupling, Glycolysis, and Markers of Beiging in Mature 3T3-L1 Adipocytes. *PLoS ONE* **2015**, *10*, e0138344. [CrossRef] [PubMed]

17. Louis, C.; Van den Daelen, C.; Tinant, G.; Bourez, S.; Thomé, J.-P.; Donnay, I.; Larondelle, Y.; Debier, C. Efficient in vitro adipocyte model of long-term lipolysis: a tool to study the behavior of lipophilic compounds. *In Vitro Cell. Dev. Biol. Anim.* **2014**, *50*, 507–518. [CrossRef]

18. Morimoto, Y.; Mori, S.; Sakai, F.; Takeuchi, S. Human induced pluripotent stem cell-derived fiber-shaped cardiac tissue on a chip. *Lab Chip* **2016**, *16*, 2295–2301. [CrossRef]

19. Morimoto, Y.; Onoe, H.; Takeuchi, S. Biohybrid robot powered by an antagonistic pair of skeletal muscle tissues. *Sci. Robot.* **2018**, *3*, eaat4440. [CrossRef]

20. Yamada, M.; Sugaya, S.; Naganuma, Y.; Seki, M. Microfluidic synthesis of chemically and physically anisotropic hydrogel microfibers for guided cell growth and networking. *Soft Matter* **2012**, *8*, 3122–3130. [CrossRef]

21. Kang, E.; Jeong, G.S.; Choi, Y.Y.; Lee, K.H.; Khademhosseini, A.; Lee, S.-H. Digitally tunable physicochemical coding of material composition and topography in continuous microfibres. *Nat. Mater.* **2011**, *10*, 877–883. [CrossRef] [PubMed]

22. Lee, K.H.; Shin, S.J.; Kim, C.-B.; Kim, J.K.; Cho, Y.W.; Chung, B.G.; Lee, S.-H. Microfluidic synthesis of pure chitosan microfibers for bio-artificial liver chip. *Lab Chip* **2010**, *10*, 1328–1334. [CrossRef] [PubMed]

23. Hwang, C.M.; Khademhosseini, A.; Park, Y.; Sun, K.; Lee, S.-H. Microfluidic chip-based fabrication of PLGA microfiber scaffolds for tissue engineering. *Langmuir* **2008**, *24*, 6845–6851. [CrossRef] [PubMed]

24. Lee, K.H.; Shin, S.J.; Park, Y.; Lee, S.-H. Synthesis of Cell-Laden Alginate Hollow Fibers Using Microfluidic Chips and Microvascularized Tissue-Engineering Applications. *Small* **2009**, *5*, 1264–1268. [CrossRef] [PubMed]

25. Zhang, Y.S.; Arneri, A.; Bersini, S.; Shin, S.R.; Zhu, K.; Goli-Malekabadi, Z.; Aleman, J.; Colosi, C.; Busignani, F.; Dell'Erba, V.; et al. Bioprinting 3D microfibrous scaffolds for engineering endothelialized myocardium and heart-on-a-chip. *Biomaterials* **2016**, *110*, 45–59. [CrossRef] [PubMed]

26. Guilherme, A.; Virbasius, J.V.; Puri, V.; Czech, M.P. Adipocyte dysfunctions linking obesity to insulin resistance and type 2 diabetes. *Nat. Rev. Mol. Cell Bio.* **2008**, *9*, 367–377. [CrossRef] [PubMed]

27. Gregor, M.F.; Hotamisligil, G.S. Adipocyte stress: the endoplasmic reticulum and metabolic disease. *J. Lipid Res.* **2007**, *48*, 1905–1914. [CrossRef] [PubMed]

MDPI

St. Alban-Anlage 66

4052 Basel

Switzerland

Tel. +41 61 683 77 34

Fax +41 61 302 89 18

www.mdpi.com

Micromachines Editorial Office

E-mail: micromachines@mdpi.com

www.mdpi.com/journal/micromachines

www.ingramcontent.com/pod-product-compliance
Lightning Source LLC
Chambersburg PA
CBHW041216220326
41597CB00033BA/5987